One-Hour Skirmish Wargames

By John Lambshead

Pen & Sword
MILITARY

First published in Great Britain in 2018
and reprinted in 2021, 2022 and 2026 by
Pen & Sword Military
An imprint of
Pen & Sword Books Ltd
Yorkshire – Philadelphia

ISBN 978 1 52670 004 9

Printed in the UK by CPI Group (UK) Ltd, Croydon, CR0 4YY.

The Publisher's authorised representative in the EU for product safety
is Authorised Rep Compliance Ltd., Ground Floor, 71 Lower Baggot
Street, Dublin D02 P593, Ireland.
www.arccompliance.com

For a complete list of Pen & Sword titles please contact

PEN & SWORD BOOKS LIMITED
47 Church Street, Barnsley, South Yorkshire, S70 2AS, England
E-mail: enquiries@pen-and-sword.co.uk
Website: www.pen-and-sword.co.uk

or

PEN AND SWORD BOOKS
1950 Lawrence Road, Havertown, PA 19083, USA
E-mail: uspen-and-sword@casematepublishers.com
Website: www.penandswordbooks.com

One-Hour
Skirmish
Wargames

*Dedicated to Shaun Murphy for his infinite patience
in playtesting my designs and just for being a mate.*

Contents

Foreword		vi
General Introduction		vii
Part I: The Early Days of Firearms		1
Chapter 1	The Core Rules	3
Chapter 2	The Age of the Musket	15
	Scenario 1: Capture the Cannon	18
Chapter 3	The Rifle Era	23
	Scenario 2: Zulu	26
Part II: The Twentieth Century		31
Chapter 4	Wars Within Peace	33
	Scenario 3: Freikorps – *'Drang Nach Osten!'*	41
Chapter 5	World War II	47
	Scenario 4: Plan Tortue	51
Chapter 6	The Cold War	57
	Scenario 5: Graveyard of Empires	60
Part III: Extending the Game		67
Chapter 7	Pulp Action	69
	Scenario 6: Retro Rockets	72
Chapter 8	Fighting a Campaign	81
Chapter 9	Points System	93
Chapter 10	Additional Rules	97

Foreword

It is both an honour but also something of a daunting task to be asked to write the follow up to Neil Thomas' hugely successful book on *One-Hour Wargames*. Neil set a high bar with his ground-breaking work that very much captured the mood of our times. Not everyone always has the time to read complicated rules or to play an all-day game on large tables involving large numbers of models.

It is worth stopping to consider what made Neil's book so successful. I would suggest the following points might be part of the explanation, other than the obvious that the rules can be quickly grasped, and the games easily played in an hour or two on a '3 by 3' table using 100 figures or less. Firstly, Neil shows that simple doesn't have to mean simplistic let alone facile; as has often been pointed out, chess is simple.

Secondly, he demonstrates that large and complicated is not synonymous with realistic. His rules cover nine eras ranging from the ancient world to World War Two, and capture the atmosphere and essence of each period by selectively tailoring a core rule philosophy. For example, the rules for the ancient period use a sequence of play reflecting the importance of the clash of spear and shield over shooting: Movement, Shooting, Hand-to-Hand Combat, and Elimination of Units. However the equivalent sequence of play for World War II rules is Movement, Observation, Shooting, and Eliminating Units, reflecting a quantum jump in firepower range and lethality.

There is a definite trend in wargaming towards slicker, faster, comprehensive, pre-packaged games that can be learnt and played in a few hours and it's clear that this is more than just a transient fad. One has only to look at the direction rules are taking at the large commercial wargaming operations like Games Workshop for confirmation of the trend. Whether it's because our world moves faster with instant and extensive communication, because of the greater demands placed upon our time, or something more subtle like the way we now take in information through screens rather than print, who knows? But, whatever, Neil's book caught the *zeitgeist* of the day.

General Introduction

In general terms, 'skirmish' in ordinary English means a short conflict which may range from an argument with one's spouse about whose turn it is to put out the rubbish to the sort of robust discussion that happens in pub car parks after last orders have been called.

In the military, the definition is rather more specific. It means a brief, unplanned encounter between small units of troops, ships or aircraft – especially by advanced or outlying detachments of larger forces.

In wargaming, a skirmish was commonly seen as a game where one model represented one man or vehicle, as opposed to 'traditional' wargames where twenty-four figures might represent a Roman cohort or Victorian rifle regiment. The simplicity of this definition is now muddied by 'company" or 'platoon' based games, including some of the most popular game systems in the hobby such as Warlord Games' *Bolt Action* or Games Workshop's *Warhammer 40K*. In such games, one model still represents a single person or vehicle BUT the models commonly move and fight in larger formations: the operational unit (OU) is not a single model but a squad of, say, ten models. An OU can be defined as the smallest number of models that move and fight independently. And when it comes to rules writing, it's the OU that actually defines the scales and mechanics of the game.

Item: a skirmish wargame is where a single model not only represents one person or vehicle BUT that one model also represents an OU.

A good rule-of-thumb is that a single gamer can handle about twelve OUs in his army; more than this and the game can bog down as players get overwhelmed, fewer than this and there are not enough OUs to allow strategies so as to give players a challenging and enjoyable game. Of course, this figure isn't immutable – very little is in game design. If the OUs all have universal, simple properties then you could easily double or treble the number that can be controlled. Conversely, if each OU is unique, with complicated, interacting special rules then even 6 or fewer OUs per army could prove challenging.

Skirmish games aren't quite as old as wargaming itself but they do have quite a respectable pedigree. The first skirmish game of which I am aware was a set of Wild West gunfighter rules by Steve Curtis, Ian Colwill, and Mike Blake that was released in the early 1970s – but no doubt there are earlier examples. 1975 saw the release of a seminal rule book just called *Skirmish Wargaming* by the redoubtable Donald Featherstone.

There is a school of thought that persists to this day that because skirmish wargaming involves few models that each model must have concomitantly complicated special rules. Often the player has been required to micromanage actions. For example a gunman model might not just shoot, but (i) locate the target, (ii) draw his pistol, (iii) cock his pistol, (iv) aim his pistol and (v) pull the trigger. The player might even be called upon to write out orders in advance detailing all these actions. This approach has meant that skirmish games have had a tendency to become complicated virtual models of real life.

I well recall playing a game of Cold War fighter combat (air games are often a sub-branch of skirmish games) in the early 80s where a single pass by two Tornados at an element of MIG 25s that might have taken ten to thirty seconds of real time actually required all afternoon to play. If game-time is longer than real-time then we're doing something wrong. In many ways the development of cheap, powerful, digital games machines with excellent graphics has made such detailed wargames superfluous. Computers are good at creating a model of a world – we call it virtual reality.

Now we can play man-to-man shoot-'em-ups and air-to-air combat in real time, leaving the hardware to cope with tedious matters such as record keeping and the mechanics of three dimensional movement, while we get on with the enjoyable task of moving and shooting. I knew the writing was on the wall when I came back from lunch to find my students had linked all the research computers together through the departmental web and were hunting each other through a metaphorical warehouse armed with virtual shotguns.

Item: a skirmish game should play in something close to 'real-time'.

And yet, and yet, there are those of us whom pixels leave cold, who like painting toy soldiers and want to field them on a table top. But we don't want record keeping, over complexity, or long drawn-out encounters. We want to recreate the feel of an action movie, the fast hit on hit of a video game, the rush of adrenaline when Clint Eastwood steps out from cover and lets rip from the hip with a Schmeisser. In short, we want excitement.

Item: Players like an heroic game so use rules that let the player concentrate on dramatic rather than bureaucratic decisions.

One way to keep a skirmish game moving at something resembling Hollywood pace is to simplify by ruthlessly focussing down on what is actually important to the player, such as fire and movement. For example, do we really need players to routinely keep track of ammunition levels for their models' guns and reload as necessary? The tedium level involved is regrettably high compared to the dramatic potential.

Item: Avoid record keeping and game 'bureaucracy' in general wherever possible.

Another way to speed up the game is to streamline the way information is presented. In particular, the game design should present all necessary information for each player on a single sheet and avoid making the players consult multiple tables especially those that involve interaction between model abilities and die-roll modifiers.

Item: Avoid the use of mental arithmetic and tables, especially those that employ ratio and modifier interactions between models.

A counter-intuitive feature of reality is that large systems featuring multiple sub-units are far more predictable than small systems with minimal sub units: it is easier to predict climate than it is weather; easier to predict air pressure than the behaviour of a single nitrogen molecule. In the same way, the result of a clash between two divisions of ten thousand troops each is more predictable than an encounter between two individual soldiers, where all might be decided on the exact trajectory of a single round.

Item: A skirmish game is more realistic with a wide spread of potential outcomes so a significant element of randomness needs to be injected.

The standard dice spread of one to six is inadequate for a skirmish game. Adding two dice rolls together does offer a wider range, from two to twelve, but the numbers now follow a normal distribution and so are centre-weighted. This weighting only gets heavier and hence the results more predictable, as more dice are used.

This problem can be addressed by dice with more than six sides, role-playing games are an obvious example. And, of course, poly-sided dice can make mental arithmetic issues more of a drag on the game. One approach to get around mental arithmetic is to use many types of poly-sided dice, D4, D6, D8, D10, D20 and so on (standard wargame convention is to abbreviate a dice numbered 1 to 6 as 'D6' and so on). Multiple types of poly-dice allow different models with different weapons and abilities to have a wide range of probabilities of success without using dice-roll modifiers simply by allocating a different dice to different models: one is more likely to roll a 6+ on a D8 than a D6, and much more likely on a D20.

But, for some reason that eludes me, wargames employing poly-sided dice mechanics never seem to be as popular as games using traditional dice, as confirmed by a quick mental flick through the best-selling games at the time of writing.

In *One Hour Skirmish Games,* I use a fairly novel approach that eschews dice altogether, replacing them with playing cards. Standard playing cards are cheap, readily available, and offer a wide variety of linearly-distributed probabilities without the game design having to resort to using number modifiers. For example:

Black or Red, probability = 50%
Suits, probability = 25%
Cards as numbers, probability = 7.7%
Individual cards, probability = 1.9%

Probability weightings between models can be easily altered by drawing more than one card and discarding all but the highest.

Aces are always low, equalling the number 1. Knaves/Jacks are worth 11, Queens 12, and Kings 13. The game uses the Bridge Convention of suite ranks when comparing cards, so Spades beat Hearts, which beat

Diamonds, which beat Clubs. Any probability resolution in the game is used by comparing the draw of a card with either a target number or a card drawn by one's opponent.

And, er, that's it really.

Item: Use playing cards rather than dice for a wide range of probabilities without using tables or mental arithmetic.

In this book I follow Neil Thomas's example of a core set of rules that are modified by special rules to capture the essence of each period. One difference is that the game is restricted to periods after firearms became dominant weapons, i.e. the adoption of muskets. Prior to this, hand-to-hand combat, the interaction between offensive muscle-powered weapons and defensive shields and armour, ruled combat. I felt that the change was so significant that identical core universal rules could not be usefully devised for both eras.

I use a convention where comments in brackets after a rule explain why the rule is shaped the way it is. Partly that's because I believe that if players know what a rule is trying to achieve it will help them understand it, but partly it's simply so players can see some of the design decisions that went into the game.

Part I

The Early Days Of Firearms

Chapter 1

The Core Rules

'All's fair in love and war, you know.'

Smedley

This chapter introduces universal core rules that apply to all the periods defined in this book, unless modified by era-specific rules.

Battlefield

A three or four-feet square playing area is recommended. It should be well cluttered by terrain to give cover for the combatants, to block lines of sight, and to channel movement. Examples might include woods, buildings, ruins, walls and machinery such as is found in an industrial complex.

Terrain affects movement, line of sight for firing and the number of cards drawn in defence against a shooting attack. It may be divided into two separate types: area and linear.

Area terrain

This may be as small as a 32mm base with a model of a tree on it or as large as an area of rubble covering, say, six inches by four inches. Area terrain must be clearly delineated by an edge and should ideally be decorated with an appropriate model(s) such as trees, so the players know exactly what the terrain is and where it is located.

Area terrain does not normally affect the movement of people on foot or animals, unless it is completely impassable, and has no effect on hand-to-hand combat. Impassable terrain includes features such as deep lakes, quicksand, large solid structures such as a giant boulder or locked building, or impossibly tangled vegetation like a giant bramble patch. Area terrain that is high enough does affect line of sight: examples might include trees. It also affects a model's defence against being shot; more on this later, in the shooting rules.

Linear terrain

These are landscape features that form lines or snake across the landscape. Examples might include trenches, gullies, streams, walls, hedges and so on. These can affect movement (depending on depth or height), and line of sight and defence against shooting attacks (if high enough). Linear obstacles have no effect on fighting hand-to-hand.

Models

Generally, models should be individually mounted, one to a base. Occasionally, bases will have more than one model for convenience. Examples include crew-served heavy weapons such as heavy machine guns or Napoleonic cannon, which are classed as a single model for the purposes of movement and firing but have two crew – we will discuss this later. I recommend using markers as reminders of knock-downs and casualties where multiple models are on one base. You can buy little plastic markers very cheaply online or from hobby shops

Cards

Each player has a pack of playing cards which is used to determine 'Actions' and 'Resolutions of Actions'.

Players turn over cards from a single deck for command control to move models, shoot, or do something specific to the scenario.

Cards are ordinary playing cards, one deck of fifty-two cards plus two jokers per player, which are used as randomizers to resolve shooting etc. Normally, card draws are opposed with each player turning over one or more cards – the highest card winning: aces are worth one point, jacks eleven points, queens twelve points, and kings thirteen points. In the event of a tie, i.e. each player draws a seven, remember that the suites have the standard Bridge Card Game rank value: **S**pades beats **H**earts beats **D**iamonds beats **C**lubs. So a seven of Spades beats a seven of

Diamonds. (A simple mnemonic to remember this order is **S**uper **H**eroes **D**on't **C**ry.)

Treat the result as a draw if both players draw exactly the same card (i.e. each draws the seven of Spades). In this case each player draws an additional card to settle the matter. If the second draw is also two identical cards then do it again. If the third draw is also two identical cards then I suggest you give up wargaming, brew a pot of tea, and work on developing The Infinite Improbability Starship Drive.

In general, probability of success is modified by drawing greater or fewer Resolution Cards. So, for example, multiple-shot weapons draw more than one card and high-quality models may draw more than one card.

Each player works his way through a deck of playing cards, drawing one off the top to reveal its value, then discarding the card after use.

The mechanism of the game is that play is divided into Turns, which are further divided into alternating Player Phases. A player phase ends when a player runs out of Action Points or elects to 'pass'. A Turn ends **immediately** either player turns over a joker.

Winning and Losing

The game ends when one player achieves a scenario objective, or a player's army breaks (withdraws from the field) or is completely destroyed. The likelihood of an army breaking depends on their Motivation, their Leadership and Casualties (that actual casualties are used rather than some percentage makes the game robust to army size and self-balancing). A Casualty might be a dead or heavily wounded soldier but, equally, it might just be a soldier who has lost his nerve and run away or simply 'gone to ground'.

The Turn

Turn Sequence

1. Each player turns over a card to check for initiative.
2. The player with the highest card has the first Player Phase.
3. His opponent then has the second Player Phase.
4. Go back to 2 until a joker is revealed.
5. Both players then check their army's Morale simultaneously. A player loses if he fails the morale check and the game ends. The game is a draw if both players fail their morale check at the end of the same Turn. If both players pass their morale check the game continues – go to 6.
6. Players then check to see which of their downed soldiers are Casualties. Draw a card for each downed infantry model to see if it is a Casualty and hence permanently out of the fight: 'red is dead' and the model is removed from play, black means all fine and the model is stood up and continues in the game.
7. Start next Turn unless the scenario rules dictate otherwise.

The Player Phase

1. The phasing player turns over a card to receive a number of Action Points. Put the card to one side as a reminder of its value. An Action Point is expended to activate a model, allowing it to move, and/or fire, and/or carry out some special scenario-specific task. A player may move a model more than once but the second move costs three Action Points and the third move costs five. A model may only fire once per Player Phase and firing ends its activation: a model can't shoot then move but it can move once, twice or three times and then shoot. The model may not move after shooting (I don't want players shooting a model then moving it into a place of safety without their opponents getting a chance to retaliate). A model may only be activated once per Player Phase and a player must complete all actions with one model before activating another.
2. A Player Phase ends when a player runs out of Action Points or because he does not wish to do anything more. Excess Action Points are lost: they cannot be carried over to another phase.

3. The second player then has his Player Phase, turning over a card as above.
4. Repeat alternating phases between players until a joker is revealed. Play **instantly** stops wherever a player happens to be in his phase and the **Turn** is over.

Note that the length of each turn is variable and unknown (this introduces the chaotic nature of skirmish games at the turn level but it tends to balance out between players over a whole game).

Downed Models And Casualties

At this point it is worth defining what is meant by the terms 'Downed' and 'Casualty' as these words will crop up throughout the rules. A 'Downed' model is one that has failed to defend against being shot and so drops to the ground (the model may have actually been shot or may simply be ducking into cover after a round comes too close). Downed models may best be represented by turning the model on its side. Add markers to crew-served weapon to indicate if either or both of the crew are Downed: the model continues to act normally until it receives two Downed markers.

'Casualty' means that the model is incapacitated and removed from play (the model may be dead, badly wounded, or have lost its nerve and run away or simply gone to ground). A crew-served weapon is removed from play when both its crew are casualties.

Infantry Combat

Movement

Infantry: 6 inches.

Infantry are mostly unaffected by terrain except for greater-than-waist-high linear obstacles that a model has to climb over, or a crevice/stream that he has to jump over or wade through. This takes a full move. Entering or leaving a standing building, as opposed to rubble, or a vehicle also takes a full move. Some linear obstacles such as deep water, obstacles greater than head-high, or thick undergrowth are impassable to infantry. Infantry may climb over scalable obstacles that are greater than head-high but this counts as two moves for the purposes of expending Action Points.

It is very important that players agree before play begins on what the terrain on the table actually represents in game terms. This will prevent many arguments and much bad feeling later.

Note that models may not move off the playing area unless specified in the scenario.

Crew-Served Weapons

Crew-served heavy weapons on bases, tripods or wheels cannot be moved from position to position during the course of a game unless allowed by special scenario rules but they may be turned in place to point in any direction. This is a move and all the usual rules apply.

All crew-served heavy weapons have two-man crews for the purposes of the game; the number of crew actually modelled is irrelevant (this is a simplification to speed the game up and prevent heavy weapons dominating but players are at liberty to change this rule to use the actual number of crew models if they wish). Note that a crew-served model removed from the table counts as two casualties when checking whether an army breaks (if players choose to use the actual number of crew operating the weapon then this number is used instead, which might be something of a liability).

Crew-served weapons may turn and fire normally as long as one crew member is not Downed or a Casualty.

Close Combat

Close combat is immediately resolved when a model moves to touch an enemy model. The attacker draws two Resolution Cards, discarding the lower, and the defender one Resolution Card. The losing model is not Downed but is instead immediately removed from the game as a Casualty. An already Downed model that is attacked in close combat has no defence and becomes an instant Casualty without recourse to Resolution Cards (close combat is designed to be decisive).

Note that a crew-served weapon such as a machine gun is a single model for purposes of close combat. It never attacks and it turns over only one card in defence. The first Casualty received does not cause a crew-served model to be removed from the game. Move the attacker back two inches. Any subsequent Casualty result by hand-to-hand combat or shooting, on the same crew-served weapon results in it being removed from the game.

A close combat is cancelled and the attacker moved two inches back if a joker is revealed, ending the turn.

In some scenarios, the close combat mechanism may be used to capture rather than kill an enemy model. A captured model is assumed to be disarmed and must be escorted by a captor model, moving with the captor. A captor may escort or carry one captured model or one scenario-specific object. If the captor is knocked down or killed then they drop any object or unconscious captive, and a conscious captive is deemed to escape. Move it two inches away from the captor model if knocked down. The escaped captive behaves normally as a player-controlled model, starting on the next player phase.

Ranged Combat Weapons

Weapons have two basic properties: effective range and number of shots. For example, a pistol can fire one shot at up to six inches range, whereas a rifle can fire one shot at a target at any distance. The effective range of a rifle is greater than the size of the playing area.

Generic rules for some commonplace infantry weapons are given below. Note that it takes one Action Point to shoot.

Pistol: range 6 inches, one shot.
Musket: range 18 inches, one shot.
Manual Rifle range infinite, one shot.
Semi-Automatic Rifle: range infinite, one shot.
Squad Automatic Weapon: range infinite, one shot OR range 36 inches, two shots.
Sub-Machine Gun/ Machine Pistol: one shot range 18 inches, OR 2 shots range 6 inches.
Light Machine Gun: range infinite, two shots.
Tripod-Mounted Heavy Machine Gun: range infinite, three shots, crew-served.

Firing

The firing procedure is as follows.

1. The shooting player checks that he has a line of sight between the firing model and the selected target and that the target is within the range of the model's weapon. If these criteria are satisfied the player may expend an Action Point and shoot.
2. Both the shooting and the defending player turn over one Resolution Card each from their decks and compare results.
3. If the shooter wins, the target model is Downed, place it on its side or mark it in some way. A Downed model may not move or fire so there is no point in firing at an already downed model.
4. If the shooting weapon has two shots and misses on the first then the shooting player may draw another card and compare it with the originally drawn defender's card, and so on.
5. Shots 'left over' from a weapon after a target is Downed may be retargeted onto any model within three inches, whether or not the firing model has a line of sight (this is to account for overs, ricochets etc). The defender draws a new card for the second model, and so on.
6. The defender turns over two Resolution Cards (discarding the lower) if the target model is in, or immediately behind a line of, soft terrain such as vegetation, three Resolution Cards (discarding the lower two) if in, or immediately behind a line of, hard terrain such as rubble or large trees, and four Resolution Cards if in a special purpose-built fortification such as a concrete bunker.

Line of sight and Terrain

Terrain that is greater than half the height of a model blocks line of sight for firing. Two models positioned each side of such a terrain piece where neither is positioned in the terrain, or touching it in the case of linear terrain, cannot see nor shoot at each other. A figure touching a linear obstacle up to shoulder high can shoot over it, and be shot at over it. It is recommended that terrain be based so the area of the base governs the exact extent of the terrain to prevent arguments.

A model positioned so it is partly in view to the shooting model behind terrain that normally blocks line of sight, such as the corner of a building, can be targeted but counts as being 'in terrain' when shot at: use soft or hard as appropriate.

Models positioned in terrain, i.e. on the terrain base, can see out of the terrain and can be seen by models outside the terrain with line of sight.

Morale

The morale of both sides is checked simultaneously at the end of a Turn. Add up the number of Casualties a side has suffered so far in the game, but do not include Downed models, and draw a Resolution Card. Add the army Motivation as given in the scenario to the value of the card (high Motivation means that armies will tend to stay in combat longer despite losses). The total must beat the number of Casualties or the player loses. His army withdraws from the field of battle.

Leaders present on the battlefield modify this process by allowing the player to draw more than one resolution card, discarding whichever he dislikes. For example, a Leader (2) allows a player to draw three resolution cards (1+2) to check for morale. Usually the lower card(s) would be discarded but the player may wish his army to withdraw before it takes further losses if the scenario is part of a larger campaign game. Downed Leaders may not aid a break check. Only one Leader may be used for this purpose in any one turn.

Remember that the army break check is carried out before testing whether Downed models become Casualties that turn (playtesting showed that the game flowed better this way around).

Testing for Casualties

Each player checks every one of his Downed infantry models for recovery. Draw a Resolution Card for each Downed model. A black card signifies that the model is still in the fight. Stand the model upright. A red card indicates that the model is a Casualty and that it takes no further part in the game. Remove the model.

Crew-served weapons may be Downed by ranged fire just like any other model. They are removed from the game when they suffer a second Casualty result. Crew-served weapons perform normally with only one Casualty marker on them (a rule-fudge but it keeps the game flowing quickly). They count as two casualties for purposes of checking for morale when destroyed.

It is convenient to line up casualties beside the playing area to facilitate army break checks.

Leaders

Leaders have a leadership value that is expressed as additional Resolution Cards that may be drawn when testing an army's morale. This will normally be specified by the scenario but generally an army will have one Leader of leadership value 3 and a second-in-command of leadership value 1. Only an exceptional person would have a leadership higher than 3.

The leadership value is also used for the leader model to draw additional resolution cards when in close combat or firing, but not when shot at (this gives a player difficult decisions to make on how much to expose a leader to danger so as to draw on his additional combat capability). A Leader 1 gets one additional Card, Leader 2 two additional cards and so on.

Special Capabilities

Certain models may have special abilities without being Leaders. Examples are given below.

Dead Shot (x): Model draws (x) extra Resolution Card(s) when shooting. Eg A character with Dead Shot (2) draws two extra Resolution Cards.

Bruiser (x): Model draws (x) extra Resolution Card(s) in close combat in both attack and defence.

Inspiring: Once per Turn the player may discard any Resolution Card drawn for his army and draw another.

Lucky: After it is knocked down the model is only lost on the drawing of a Heart card, rather than any red card.

Fast: Model moves +3 inches.

Tough (x): Model draws (x) extra Resolution Card(s) when defending against a ranged weapon.

Scout (x): Model draws (x) extra Resolution Card(s) when defending against a ranged weapon when in cover.

Motivation

At this stage it is worth making a few points about motivation and leadership, which superficially appear to have the same function, i.e. to reduce the chance of an army breaking, but actually work in different ways.

Motivation represents how keen an army is to fight and not get discouraged. Generally speaking, highly trained and well-equipped armies have higher motivation and elite units, with a strong *esprit de corps*, the highest. However, irregular forces can be highly motivated if they are driven by some political or religious ideology that promises a heaven on earth (or shortly after death) or if they are simply motivated by extreme hatred.

Well-trained and equipped armies can have low motivation depending on the circumstances. Mercenaries, for example, are notoriously disinclined to take heavy losses as it's bad for business to lose half one's assets merely for a victory in someone else's war. A veteran unit might have low motivation because it knows this particular battle is in itself irrelevant, a feint or a delaying action, or a veteran unit may be completely burnt out through over-commitment. So motivation in any specific unit may vary quite considerably from battle to battle.

The point about motivation is that it is a general property of the unit, not subject to the death or survival of a specific model.

Leadership represents the ability of some individuals to inspire trust and loyalty such that people will follow them into hell and make sacrifices that they would not normally countenance. The downside of an inspirational leader is that their death can have an immediate negative impact on a unit's morale (it can work the other way but rarely).

These fundamental differences are reflected in the mechanisms used to reflect general motivation and leadership in this game. Motivation is fixed and completely protects an army from breaking until losses go over a critical limit. After this point, it still always acts as a modifier to the draw of a card for a break test but at a fixed value so the more casualties mount the less motivation affects the probability of breaking.

Leadership never completely protects an army from breaking but always reduces the probability of an army breaking whatever the current number of casualties. For example, a Leader (1) halves the probability of breaking and a Leader (3) reduces it by 75 per cent, and these probabilities are fixed irrespective of the number of casualties. However, this protection only lasts as long as the leader is upright and functional. Given that leaders are also powerful attacking game pieces in their own right this gives players a difficult decision to make. A leader at the front of an attack in the position of greatest danger is inspirational for the led. So do you commit a leader in a dangerous 'boom or bust' attempt to force a decisive decision? Well, it's up to your judgement and whether you feel lucky. Note that leaders get no extra defensive ability as a deliberate design mechanism to make the point.

Chapter 2

Age of the Musket

'At Blenheim and Ramillies, fops would confess, they were pierced to the heart by the charms of Brown Bess.'

Kipling

Introduction

Muskets were the first firearms to become the primary weapon of armies. They were smoothbore, muzzle-loading, black-powder devices. The classic British 'Brown Bess' musket of the Napoleonic Wars was a flintlock weapon firing a large ball that was smaller than the barrel diameter to avoid problems caused by black powder fouling the bore. This reduced accuracy to about 50 metres, or perhaps 80 metres in the hands of an experienced soldier, when fired at a man-sized target. A trained man could get off as many as four rounds a minute and the weapon was fitted with a long bayonet, although it also made a decent club as it weighed nearly 5kg.

Muzzle-loading rifles, such as the famous British Army Baker Rifle, were the second important infantry firearm of the period. Their main feature was a rifled barrel necessitating the ball being wrapped in greased cloth or leather to engage the rifling. This greatly increased the accuracy of the weapon compared to a musket. Hitting a man-sized target at 200 metres was easily possible. Rifleman Plunket memorably shot a French general, and his aide-de-camp, at a range of around 500 metres. The downside was that forcing the ball down the rifled barrel halved the rate of fire compared to a musket. So rifles were generally only issued to elite troops who could be relied upon to exploit the weapon's range.

Officers would be armed with a sword, and possibly a pistol.

Light cavalry would be more likely to be involved in skirmish combat than their heavier colleagues as their duties included reconnaissance and outpost protection of the main force. They could be equipped with carbines (similar to muskets) that would be fired while stationary at about 100 metres. At close quarters light cavalry would use a pistol (or two) before attacking with a sabre. Lancers could also be armed with any, or all, of these weapons as well as their lance. The French in particular made heavy use of dragoons that could fight as cavalry with sabres and pistols, or as infantry with muskets or musketoons (a carbine/shotgun dual weapon).

Napoleonic Special Rules

Infantry Weapons

Muzzle-loading Rifles: Two action points rather than one must be expended to shoot a muzzle-loading rifle (this rule covers the extra time taken to load such a rifle compared to a smoothbore weapon).

Carbines & Musketoons: For game purposes, treat these simply as muskets (the difference in capability of these weapons from a musket is too small to merit cluttering the game with extra rules).

Cavalry Rules

Cavalry Carbines: A cavalryman must dismount to use a carbine at full range. If fired from horseback, a carbine has a range of 12 inches.

Cavalry Movement: Cavalry move 9 inches, or 6 inches in terrain. They are subject to the normal rules for impassable terrain and linear obstacles but cannot climb.

Cavalry Dismounting: A cavalryman may dismount but this counts as a move so requires an Action Point (or three or five points respectively if it is a second or third action for that model). Remove the cavalry model and replace it with an appropriate foot model. Dismounted cavalry may not remount for the rest of the game. For simplicity we may assume the horse very sensibly heads for safer parts (though it would be simple to add your own rules for horse holders or tying horses to a tree, hitching rail etc).

Cavalry In Close Combat: Draw one extra Resolution Card (to account for the advantage of height when on horseback).

Cavalry Lances: Draw one extra resolution card in hand-to-hand combat in addition to the extra cavalry card, ie three in total (lances are especially dangerous because of their reach).

Cavalry Defending In Terrain: Do not receive the additional resolution cards for being in hard or soft terrain when defending against a ranged weapon (horses are just too visible).

Artillery Rules

Cannon: range infinite, three shots (it is assumed to be firing grape-shot), but may only fire at a target that is within a 45° front arc of the muzzle. An artillery cannon is a crew-served weapon which requires three action points to load and fire.

(Although the cannon draws three resolution cards when it fires, they are all aimed at a single target. I assume that skirmishers are well spread out when faced with cannon fire and that the range is too short for grapeshot to spread wildly. Also the cannon cannot be traversed while firing as can a machine gun. However, players who disagree with this ruling can introduce the richochet rules given later in this book to simulate the spread of grapeshot – as opposed to a single giant ball.)

Alternative Armies

These rules will work equally well for skirmishes set in the War of the Spanish Succession, The Golden Age of Piracy, Border Reaver encounters or, at a pinch, the English Civil War – or even clashes between buccaneers and Spanish troops.

Scenario 1

Capture The Cannon

Introduction

Late October, 1810, a French Army commanded by Masséna has been slowly starving while dug in besieging the Anglo-Portuguese Lines defending Lisbon. After a month the General decides on a tactical retreat north east towards Santarém, pursued by Wellington's army. In the retreat a French cannon is abandoned by its crew. Aggressively commanding the French rearguard, a furious Ney sends back a small detachment to secure the gun.

The detachment arrives just in time to see a British advanced patrol surrounding the gun. Leaving the draft horses and drivers behind, the Napoleonic light infantry and cavalry launch an attack to drive off the British.

The Battlefield

The countryside is grassy, with hills, rocky outcrops and small stands of trees. There may be farm buildings or ruins. The 'road' is a pebble-and-dirt-surfaced lane.

The British set up first anywhere on their half of the table.

The French set up second anywhere within six inches of their table edge.

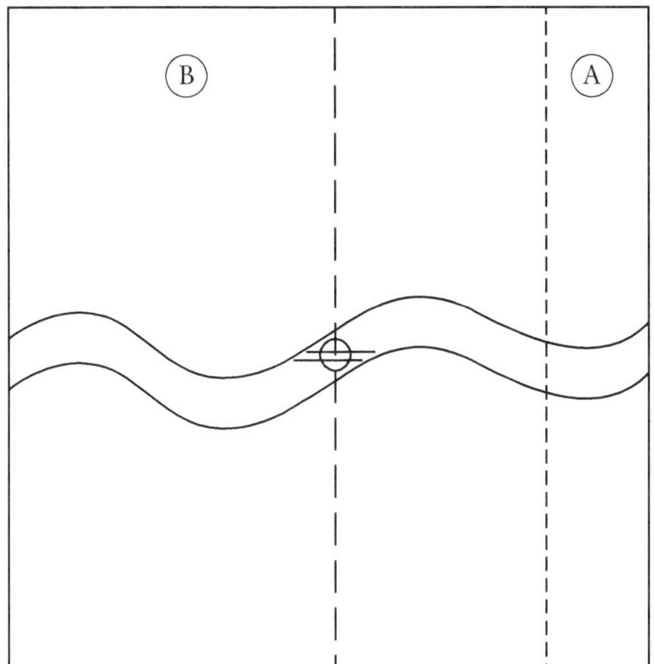
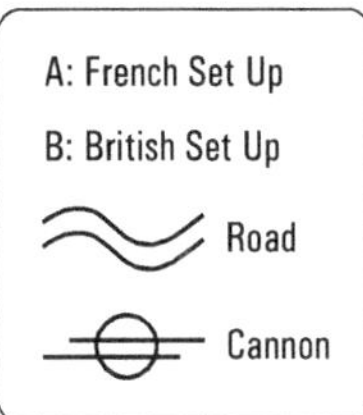

Game Length

The game ends after five turns or when one or both of the armies are broken.

Objective

French: Drive your opponent from the field of battle so you can recover the gun.

British: Stop the French recovering the gun.

Winning: The British win if they have a soldier within 6 inches of the gun at the end of the game. The French win if they break the British army. Any other result is a draw.

Special Rules

None.

Players Notes

The British player wins simply by surviving and holding his position so the French player has it all to do. French cavalry is particularly valuable for charging into hand-to-hand combat; don't waste them in a firefight with riflemen that they'll lose.

Forces

British Army (26 points)

Motivation: 3 (well supplied and on the advance)

58th Regiment of Foot (The Rutlands)
Corporal Riley, musket, Leader (1)

Private O'Gooley, musket, Bruiser (1)

5 x Privates, musket

60th Regiment of Foot (The Royal American Rifles)
Sergeant Flat, Leader (2), rifle

Private Fenman, rifle, Sniper

2 x riflemen, rifles

Weapons
Musket: Range 18 inches, one shot
Muzzle-Loading Rifle: range infinite, one shot (needs two action points to fire)

Special Capabilities
Sniper: Fenman is tasked with killing French officers so may draw 2 extra
Resolution Cards when firing at Lieutenant Griois
Bruiser (1): Model draws an extra Resolution Card in close combat in both
attack and defence

French Army (32 points)

Motivation: 2 (half-starved and retreating but on the other hand who wants
to report failure to General Ney?)

Chasseurs-à-Cheval
Lieutenant Griois, cavalry, pistol, Leader (2)

5 x chasseurs, cavalry, carbine

Voltigeurs
Sergeant Talhouet, musket, Leader (1)

9 x privates, muskets

Weapons
Pistol: range 6 inches, one shot.

Musket/Carbine: range 18 inches, one shot.

Carbine fired from horseback: range 12 inches, one shot.

Chapter 3

The Rifle Era

'The sand of the desert is sodden red, Red with the wreck of a square that broke; The Gatling's jammed and the Colonel dead, And the regiment blind with dust and smoke.'

Newbolt

Introduction

The second half of the nineteenth century saw the rise of breech-loading weapons that incorporated 'firearm actions' originally invented by American Jacob Snider, who died in London in poverty while awaiting payment from the British Government. The development that made breech-loaders effective was the metallic cartridge. The first purpose-designed breech loader used by the British Army was the lever-action Martini-Henry of 1871. The effective range of the weapon against a man-sized target was 300-400 metres. It was in service until after World War I and 'Pass-Made' copies of Martini-Henrys still turn up in Taliban arms caches today.

The most famous use of the Martini-Henry was at Rorke's Drift in the Zulu War. The rifle was phased out after 1888 when the first bolt-action, detachable-magazine rifles went into service in the form of the Lee-Metford and, eventually, the Lee-Enfield. In a number of colonial wars it became clear that spear and shield armed warriors had no particular advantage over rifle and bayonet armed European troops even in close combat.

Multiple-shot weapons in the form of volley guns or revolvers are almost as old as the firearm itself, but the first practical controlled multi-fire weapon with mechanical loading was arguably Richard Gatling's gun of 1861. Competing workable, hand-cranked designs included the

Gardner and the Nordenfelt (incidentally, it was a Gardner not a Gatling that jammed at the Battle of Abu Klea, as described above).

Artillery evolution in the second half of the nineteenth century paralleled rifle development. Armstrong produced 64mm light field guns with breech loading and rifle barrels, while the French '75, with the additional advantage of a hydro-pneumatic recoil mechanism, was arguably the first modern cannon. Indeed, because the trail and wheels remained stationary, the '75 had a higher rate of fire than contemporary rifles. Explosive shells are as old as artillery itself but in this period were not particularly effective at a skirmish level.

European armies were armed and trained to fight other similarly equipped European forces but the British, in particular, often found themselves fighting asymmetrical combats against regional powers with semi-medieval or even iron-age weapons and tactics. The most famous examples of these 'Warrior Armies' were the Zulu and the Sudanese.

The Zulu were traditionally armed much like ancient soldiers with a shield and stabbing spear. They fought in loose formation, rather than a shield wall, so used the spear underarm like a sword for maximum effect. In terms of hand to hand combat, the Zulu appear to have had no particular advantage over a British soldier using a bayonet or his rifle butt. They made a limited use of javelins. By the time of the Anglo-Zulu War of 1879 at least half the Zulu army would have had access to a firearm, but most would have been muskets.

Zulu standard tactics in both skirmishes and large-scale battles was to close fast and envelop the enemy on each flank, settling matters with a ferocious charge. They were highly motivated and would press home attacks with great courage.

Special Rules

Warrior Rules

Massed Charge: The player running a warrior army turns over two cards, discarding the lowest value, when checking for Action Points (this is to simulate the eagerness of warriors to engage the foe).

Fast Attack: Warriors armed only with spears/swords and shields move as cavalry rather than infantry (the British Army tactically treated Sudanese warriors charging as cavalry because of their speed).

Advanced Weapons: Warriors occasionally acquired modern firearms (usually after a victory over a modern army) but tended to make poor use of them, filing off the sights and other bad maintenance practices, so treat all such weapons as muskets for game purposes.

Hand Cranked Multi-Shot Weapon (Gatlings etc) Rules

Multi-Shot Gun: range infinite, three separate shots. Turn a resolution card over for each shot one at a time until either all have failed to overcome the target's defensive resolution cards or the target is downed.

Ricochet: Shots 'left over' from a weapon after a target is downed may be retargeted onto any model with three inches, whether or not the firing model has a line of sight (this is to account for overs, ricochets etc.).

Jamming: The player may choose to fire one, two or three shots with these weapons as there is a chance of them jamming. Jamming happens if two or more resolution cards of the same suite are turned over when firing. The gun fires normally, then jams. In a following phase the player must expend two action points to unjam the weapon, whereupon it behaves normally.

Crew-served: The usual crew-served rules apply.

Alternative Armies

These rules may be used for the many asymmetric 'European Army versus Warrior Army' wars set in Africa or other parts of the world. They may also be used for clashes between 'European' armies such as the American Civil War, Franco-Prussian War, Third Carlist War, Russo-Turkish War, and Boer Wars. The American Cattle Baron Range Wars and the 'Wild West' in general offer much skirmish wargaming potential.

Scenario 2

Zulu

Introduction

The Zulu army that had been so victorious at the Battle of Isandlwana suffered a heavy defeat at Khambala, a battle that was essentially Rorke's Drift writ large, with the Zulus attacking a fortified British position. The Zulu king, Cetshwayo, had given standing instruction to his forces to refrain from attacking fortified camps and to try to tempt the British into the open but it was not that simple. The Zulus had a warrior culture and were easily provoked into a charge against their foe whatever the tactical situation.

Chelmsford launched a second invasion of Zululand in May, 1879. The Zulus resorted to hit-and-run guerrilla tactics against the extended British supply chain, forcing Chelmsford to build a series of forts along the route of the crawling wagon trains. Prince Imperial Louis Napoléon died in June in just such a skirmish.

This scenario depicts an encounter between a Zulu *impi* (warband) of the uVE ibutho (regiment), and a small British detachment from the 90th Perthshire Light Infantry manning a redoubt where the supply trail crossed a stream running into the River Mhiatuze. The lieutenant in charge despatched a mounted messenger back down the trail to summon help from the closest fort but his orders demand that he holds the position until relieved. The Zulus receive a continuous stream of reinforcements.

The Battlefield

The terrain consists of flat rolling grassland, the only cover consisting of small outcrops of rock and giant ant hills. The stream demarks the eastern edge of the battlefield and is a linear obstacle except at the ford. The redoubt should be about 8 x 6 inches.

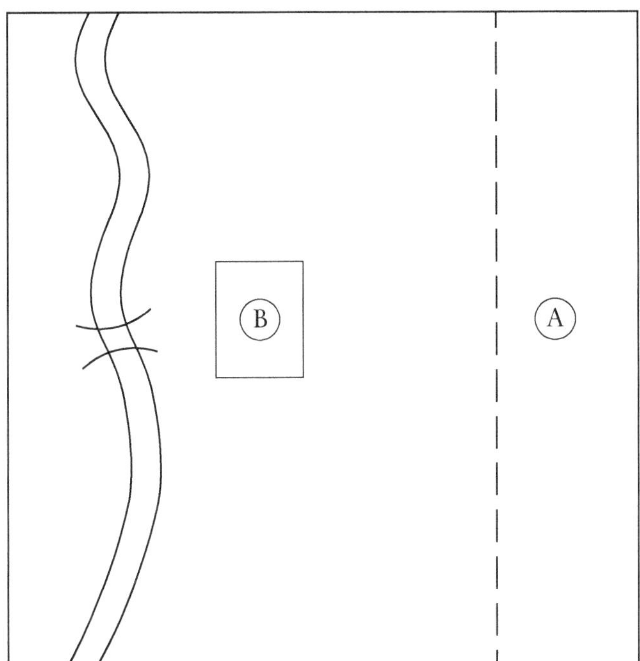

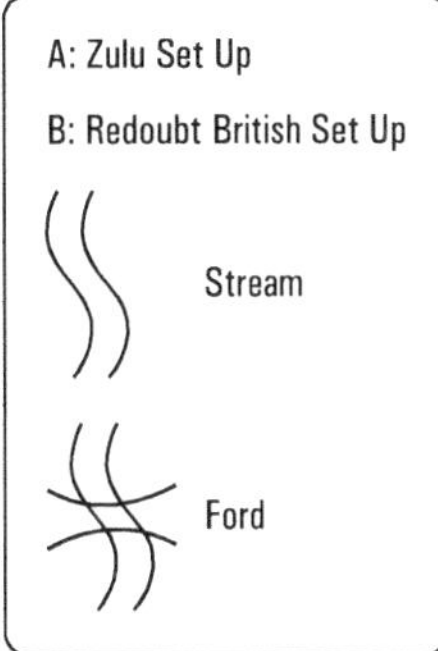

The British set up in the redoubt and the Zulu force at least 18 inches from the redoubt.

Game Length

The game ends when British reinforcements arrive or one of the armies is broken.

Objective

Zulu: Destroy the invaders.

British: Hang on until relieved.

Winning: The British win if they have at least one undowned soldier within the redoubt at the end of the game provided (i) the British army is not broken, and (ii) there are no undowned Zulus within the redoubt. The Zulus win if they have at least one undowned soldier within the redoubt at the end of a turn and there are no undowned British within the redoubt. Any other result is a draw.

Special Rules

The Redoubt

Cover: A British soldier inside the redoubt receives +2 Resolution Cards when fired at from outside.

Improved Position: A British soldier inside the redoubt receives +1 Resolution Cards defending a wall when attacked by a Zulu Warrior outside the redoubt.

Zone of Control: British soldiers guarding the wall exert a 3″ zone of control along it (i.e. 1½″ to each side of where their round base touches the wall. A Zulu cannot cross a wall within that zone of control. Downed soldiers have no zone of control.

Reinforcements

Zulu: Warriors removed as casualties are brought back onto the battlefield positioned at least 12 inches from the redoubt as reinforcements. They spend one turn off the table then come back at the start of the following turn and may be used as normal. The Zulu player will have to keep count of their total casualties as the same model may become a casualty more than once. Models returning as reinforcements retain their weapons and special capabilities. Exception – a leader when lost comes back with his leadership value reduced by 1 (if initially greater than 1).

British: after checking army morale the British player draws a Resolution Card and notes the value. These are added together after each turn. British reinforcements arrive, ending the game when the total of this addition exceeds 30.

Players Notes

The Zulu *induna* (commander) is absolutely critical to keeping the Zulu army fighting. If he is lost, the army becomes susceptible to being broken by a single unlucky card. The British player should strive to keep Zulu warriors out of the redoubt and kill any who enter as a priority. The extra defence cards when defending the walls are vital for the chances of the British holding out.

Forces

British Army (18 Points)

Motivation: 3

Lieutenant Rennie, pistol, Leadership (2)

Sergeant Moynihan, rifle, Leadership (1)

Private Alexander, rifle, Inspiring

5 x riflemen, Rifles

Weapons
Pistol: range 6 inches, one shot.

Manual Rifle: range infinite, one shot.

Special Capabilities
Inspiring: Once per Turn the player may discard any Resolution Card drawn for his army and draw another.

Zulu Army (31 pts)

Motivation 5

Induna Nonkululeko, spear and shield, Leadership (3)

Mandla, spear, Bruiser (2)

Thalente, musket, Scout (1)

10 x warriors, spear and shield

3 x warriors, musket

Weapons
Spear: No special rules

Musket: Musket: Range 18 inches, one shot

Special Capabilities

Bruiser (2): Model draws two extra Resolution Cards in close combat in both attack and defence

Scout (1): Model draws an extra Resolution Card when defending against a ranged weapon when in cover

Part II

The Twentieth Century

Chapter 4

Wars Within Peace

'War had taken hold of them and would never let them go. They would never really belong to their homes again. The war was over ... but the armies were still in being.'

von Salomon

Introduction

In 1884, Albert Vickers financed the new Maxim Gun Company to manufacture Hiram Maxim's new weapon of mass destruction. This revolutionary device used recoil to mechanically eject the used cartridge, cock the gun, and reload it. The weapon pumped out ten rounds a second and was astonishingly reliable. Famously, in August 1916 a company from the Machine Gun Corps operated their ten Vickers guns continuously for twelve hours, firing a million rounds and wearing out 100 barrels – all without a single jam.

Machine guns were first used in colonial warfare with devastating results. Seven hundred soldiers fought off 5,000 Matabele warriors at Shangani, with just four Maxim guns. At Omdurman, 1898, quick-firing cannon and Maxims slaughtered the world's last great medieval army. But it was not long before conflict happened between European armies where both sides had the Maxim gun, and the battlefield became empty as soldiers were driven below ground to survive.

World War I spurred the development of light skirmish weapons that could be carried by attacking infantry, notably light machine guns such as the Lewis, sub-machine guns like the Bergmann MP18, hand grenades, flame throwers like the Flammenwerfer M.16, and light, infantry-operated, high-trajectory bomb throwers such as catapults, crossbows and light mortars.

The need to transport men and guns across the hellish artillery-and-Maxim killing zone inspired the development of armoured vehicles – first armoured cars and then finally tracked tanks and armoured transports. These spurred the development of specialized anti-tank weapons for the infantry such as the Mauser 13mm anti-tank rifle and improvized anti-tank grenades.

Much of World War I was ghastly industrialized slaughter, with little of interest for skirmish wargaming, but there are potential scenarios in trench raids in the West or manoeuvre on the Eastern Front. However, there is a great deal of scope in the 'Wars in Peace' that happened in the interregnum between 1918 and 1939. World War I, the first round, destabilized many of the states of Europe. The Russian, German and Austro-Hungarian Empires collapsed leading to civil war and constant conflict among the putative successor states for possession of territory. In the West, both France and Britain came close to revolution and civil war: tanks were used to put down unrest in Glasgow, and Southern Ireland successfully revolted and seceded from the United Kingdom – only to enter a state of civil war itself.

Special Rules

Infantry

Anti-tank Rifle: range 18 inches, one shot, armour piercing 1.

Grenade: range 9 inches, one shot, blast radius 2", armour piercing 1.

Tripod-Mounted Heavy Machine Gun: range infinite, three shots, crew-served.

Note: only one 'to hit' Resolution Card is drawn when using a grenade, not two. A model will usually be armed with a gun as well as grenades but can only fire the gun OR throw the grenade once per phase.

Artillery

The artillery rules are adjusted for this era to take into account high explosive (HE) shells. Firing the weapon takes just one action point. Test to see if the weapon hits the model and causes a knockdown as for any single shot weapon but draw two Resolution Cards rather than one. If the round misses no further action is taken. If it hits and causes a knockdown then test for every other model within the blast radius of the original target, drawing one Resolution Card for each. Blast radius is measured from the edge of the target model's base (or hull if a vehicle) and affects any model whose base/hull it touches.

Artillery Shield: a metal shield on an artillery gun confers a +2 cards defence on the crew; this replaces any possible terrain defence bonus rather than adding to it.

Mortar: Range infinite, minimum range 10 inches, one shot, blast radius 2 inches, armour piercing 1

Field Gun: range infinite, minimum range 20 inches, one shot, blast radius 3 inches, armour piercing 3

Portable Flamethrowers

These are truly nasty weapons, so I have hobbled them as much as is reasonable.

Flamethrowers: range 6″, automatic knockdown on target model, plus draw one card for every model within three inches of target: the defenders only get one card for defence irrespective of any terrain they are in.

Portable flamethrowers have no effect on vehicles.

The infantryman carrying the flamethrower only moves 4″ as it is bulky.

Vehicles

Motivation

Vehicles normally have the same motivation value as the rest of their army and are part of that army, but it can be different for special scenarios.

Movement

Wheeled Motor Vehicles: 6 inches off road, 24 inches on road. These include motor bikes, lorries, cars and armoured cars. They may not move across any sort of linear obstacle, deep snow, water or marshy ground.
Tracked Vehicles: 12 inches.

Some terrain will be impassable to even tracked vehicles. Examples might include ruins, heavy walls, deep water, large trees and so on.

Vehicles must end up facing the direction in which they have predominantly moved unless they reverse, which is at half speed. A vehicle can turn on the spot to face any direction as a move.

Transports

Many vehicles can transport passengers. To enter a vehicle, move the model up to the vehicle and then make another move to climb aboard. Place the model to one side. The passenger model cannot be activated while it is a passenger. To leave a vehicle, activate the model and make a move which places the model beside the vehicle. It may make second or third turns or fire using the normal rules.

Armour

A vehicle without armour may be shot at by any weapon. A vehicle with armour of value 1 or more may only be shot at by weapons that have an armour piercing capability.

Close Combat

Vehicles are highly vulnerable to close assault by infantry. This is both realistic, and a deliberate game mechanic. Any player who lets enemy infantry within touching distance of his vehicles deserves everything he gets. Any infantryman can attack a vehicle he is touching, including armoured vehicles at Armour Piercing (1).

Shooting at Vehicles

A vehicle is only completely destroyed when it is abandoned by its crew. Some crews will abandon their vehicle at the slightest damage while others will fight on until it is shot to pieces under them (there are examples of Soviet tankers abandoning their moving vehicle after pointing them at the enemy line while other wounded crew opened fire from their tank days after it had been supposedly knocked out). Crew's morale is tested for every time the vehicle receives damage. Damage to a vehicle is accumulative over the course of a game.

A vehicle performs normally until it is abandoned. An abandoned vehicle is considered 'dead' but is left on the battlefield as a line of sight and movement obstacle. The crew from an abandoned vehicle count as 'Casualties' for purposes of morale. They take no further part in the battle. A generic tank or self-propelled gun has four crew, generic armoured cars or light tanks three crew, generic armoured carriers two crew, while generic lorries and cars have one crew.

The situation is slightly different for passengers being carried inside a lorry or armoured carrier that is abandoned by its crew. They are placed in the Down position around the abandoned vehicle and tested at the end of the turn as normal for Casualties.

Only ranged weapons with an armour piercing number can damage armoured vehicles. The procedure for firing a weapon at a vehicle is as follows:

1. Check range and line of sight, if the target vehicle is armoured and whether the weapon is armour piercing. Both players turn over a Resolution Card to check if the vehicle is hit.
2. Assuming a hit, both players turn over a second Resolution Card. The attacker adds the armour piercing value of the weapon to the numeric value of his card. The defender adds the armour value of his vehicle and the vehicle motivation to the numeric value of his card.
3. Compare the totals. If the attacker's total is more than the defenders then the difference is the number of accumulative damage points inflicted on the vehicle.

4. Test immediately for crew morale if damage has been inflicted. The vehicle's owner turns over a Resolution Card and compares its numeric value against the current accumulated damage that has been sustained by the vehicle. The crew bail out if the numeric value of the Resolution Card is less than or equal to the accumulated damage and are immediately counted as Casualties.

Generic Armour-Piercing Weapons
Anti-Tank Rifle: range 18 inches, one shot, Armour Piercing 1

Heavy Machine Gun with AP rounds: range infinite, three shots, Armour Piercing 1, Ricochet

Bazooka/RPG/PIAT: range 30 inches, one shot, Armour Piercing 4

Automatic Cannon: range infinite, two shots – one extra shot per additional barrel if it is a twin or quad-barrelled weapon; Armour Piercing 2, Ricochet

Light Tank Gun such as the British 2pdr: range infinite, one shot, Armour Piercing 3

Medium Tank Gun such as the British 6pdr/Allied 75mm: range infinite, one shot, Armour Piercing 5, blast radius 2 inches for 6pdr or 3 inches for 75mm

Heavy Tank Gun such as the long-barrelled German 75mm/American 77mm: range infinite, one shot, Armour Piercing 7, blast radius 3 inches

Tank-killer such as the British 17pdr or German '88: range infinite, one shot, Armour Piercing 9, blast radius 3 inches
Normally only one side will have an armoured vehicle as this is an infantry skirmish game, but players may use these armour piercing values for vehicle mounted guns when firing at another armoured vehicle if a special scenario has armoured fighting vehicle(s) on both sides.

Cannon firing high explosive shells have a minimum range of 20", the same as field guns.

Vehicle Armour Value – Front/Side/Rear
Draw lines from the middle of the vehicle diagonally out through each corner of the hull to get the four 'facings' of front, right or left, and rear. The location of the firer within these facings governs what armour value is used when a vehicle is fired upon.

Unarmoured Vehicle or Tachanka: 0/0/0

Armoured Carrier/Car such as a Hanomag or WWI tank: 1/1/1

Light Tank such as a BT7: 2/1/1

Medium Tank such as a Sherman: 4/2/1

Heavy Tank such as a Tiger: 6/3/2

Note: Assault guns tend to have similar or better armour than the equivalent tank while self-propelled guns have armour similar to an armoured carrier.

Vehicle Light Weapons
Vehicle Mounted Machine Gun: Range infinite, three shots, Armour Piercing 1 Ricochet.

Note: If the vehicle has twin, linked identical turret weapons such as two MGs then both can be fired, at the same target, but if it has two different weapons such as a cannon and auxiliary then only one may be fired per phase.

Vehicle Flamethrowers: Let's not go there. It was not unknown for infantry platoons to surrender when faced with one of these horrors. Heavy flamethrowers have no place in a skirmish game.

Special Crew Capabilities
Heroic Commander (x): Model draws (x) extra Resolution Card(s), discarding the lowest, when testing to see if crew abandon vehicle

Ace Gunner (x): Model draws (x) extra Resolution Card(s) when firing the vehicle's primary weapon

Veteran Driver: May move an extra 3 inches

Alternative Armies

These rules can be used for the mobile theatres of World War I such as the early days on the Western Front, Africa, the Middle East and the Eastern Front. They are most suitable for the various insurrections and border 'incidents' that took place between the two World Wars, notably the Russian Civil War and eastern wars of independence, Finnish Civil War, German Spartacist Uprising, Turkish wars of independence, Soviet-Polish War, Third Anglo-Afghan War, Rif War, Irish War of Independence and Civil War, Chinese Civil War and Sino-Soviet Wars, Sino-Japanese Wars, Chaco War, Austrian Civil War, Italo-Ethiopian Wars, Arab Palestine Revolt, Spanish Civil War, Soviet-Japanese Wars, and many, many others.

Scenario 3

Freikorps – *'Drang nach Osten!'*

Introduction

The Freikorps were private militias to supplement the official, treaty-limited army; they were raised by German military officers or ex-officers and funded by the Weimar Republic. The purpose was to provide Germany with the means to fight Soviet-backed German communist insurgents such as the Sparticists, and communists and nationalists in countries bordering Germany. Regular military officers, such as Guderian, rotated in and out of Freikorps. The ranks were often formed from WWI veterans, especially Stormtroopers. There is a clear thread from the Stormtroopers to the Freikorps, SA and to the SS. The Freikorps were disbanded in 1933 and many of their leaders purged in The Night of the Long Knives in '34.

The Allied Control Commission instructed German units located in the Baltic at the end of WWI to stay and block the advance of the Red Army. In 1919, some, with volunteers from Germany, were formed into the Freikorps Eiserne (Iron) Division. The Eiserne defeated the Soviets and captured Riga, saving Latvia.

However, the division's appalling behaviour and attempts to create a Germanic Empire in the Baltics alienated the locals who defeated them at Cēsis in the summer of 1919. The Eiserne Division was rebranded as the West Russian Army, ostensibly a White Russian force to fight the Soviets, but was eventually expelled by the Latvian Army.

The scenario takes place in the Spring of 1919 in a small village called Bērzbeķe, near Riga. Local villagers have hidden a valuable religious icon reputed to be the Holy Golden Cross of St Meinhard in the village before fleeing the approaching Red Army. Kapitän zur Flieg von Krumen of the Death Skulls Company of the Eiserne Freikorps is tipped off to the existence of the treasure by his Russian mistress, the Countess Ludmilla. The Death Skulls march to Bērzbeķe to 'liberate' the cross, only to find the Bolsheviks already in occupation.

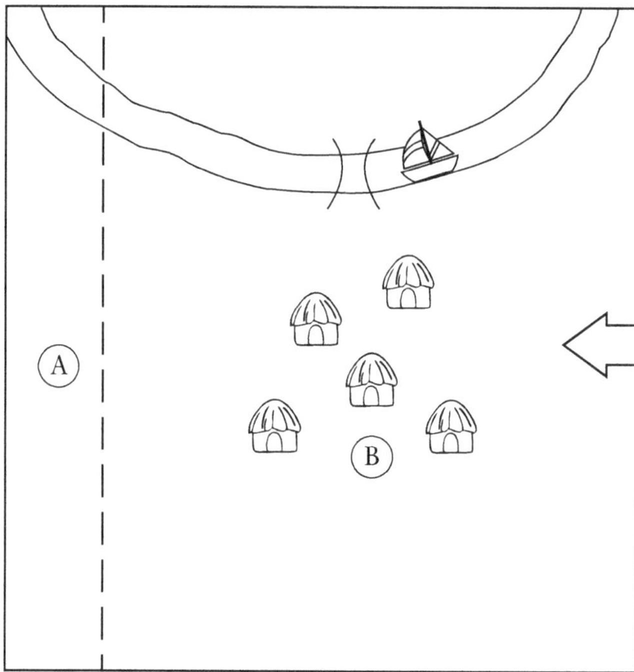
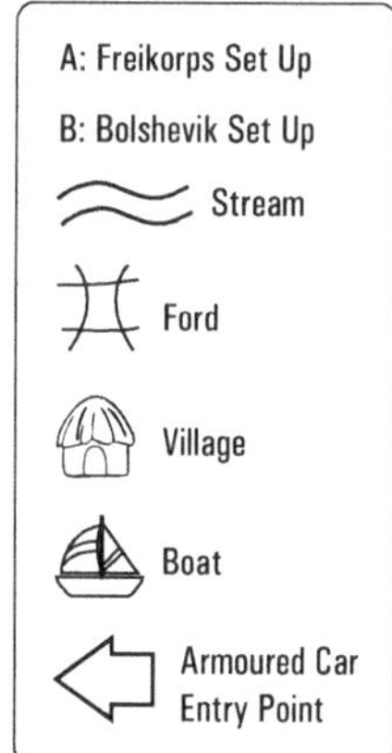

The Battlefield

Use flat north European terrain with a village consisting of suitable rural buildings at one end. Place five objective markers in different buildings making up the village and its surrounding outhouses. By all means be creative as to what constitutes a building: a model of a boat tied up on a river bank or a bridge might suffice. Use trees, animal enclosures and low hills to break up line of sight.

The Freikorps start the game on the western table edge.

The Bolsheviks start the game set up anywhere in the village except for the armoured car.

Game Length

The game lasts for six turns or until one or more of the armies break.

Objectives

The Freikorps player wins the game by finding the Holy Golden Cross of St Meinhard.

The Bolshevik player wins if the Freikorps player fails to find the Cross.

A draw only occurs if both armies break on the same turn.

Special Rules

To find the Holy Cross, the Freikorps player moves a model into a building with an objective that contains no non-downed Bolsheviks. The model may search the building by paying three action points and drawing a card. If a Queen or King is revealed the model has found the Holy Golden Cross of St Meinhard and the game ends immediately in a Freikorps victory. Any other result means that the Cross is not present: remove the objective marker. Once four markers have been removed, the fifth must be the Holy Cross. The Freikorps player therefore passes the search test automatically in this situation. Note that the Freikorps player also wins if the Bolshevik army is broken and flees as the Cross may then be found at the Freikorps player's leisure

The Bolshevik player tests at the start of each turn to see if the armoured car enters the board on the eastern table edge. He draws one card on the first turn, two on the second and so on. The car enters when a Queen or King is drawn.

Players Notes

The Bolshevik player must feed soldiers in to buildings with objectives to prevent the Freikorps player from searching them. Ideally, one needs two Bolsheviks per objective building in case one of them is downed. It is tempting to pile all troops into the objective buildings but there is often an advantage in defending the edge of the village to keep your opponent's models well clear of the objectives.

The Freikorps player must get stuck in as fast as possible. You only have six turns – and you can hear an armoured car approaching in the distance.

Forces

Freikorps (35 points)

Motivation: 3

Kapitän zur Flieg von Krumen, pistol, Leadership (4)

Countess Ludmilla (von Krumen's White Russian mistress), pistol, Leadership (1)

Crazy Adolph, SMG, grenades, Grenade Expert

2 x Stormtroopers, SMG, grenades

1 x Stormtrooper, Lewis gun

1 x Stormtrooper, AT rifle

8 x Stormtroopers, rifles

Weapons
Pistol: range 6 inches, one shot

Rifle: range infinite, one shot

Lewis gun: range infinite, two shots

SMG: one shot range 18 inches, OR 2 shots range 6 inches

Grenade: range 9 inches, one shot, blast radius 1"

Anti-tank rifle: range 18 inches, one shot, armour piercing 1

Special Rules
Grenade Expert: Draws two Resolution Cards rather than one to test to hit the target

Bolsheviks (30 points)

Motivation: 2

Captain Yudenich, Military Specialist, Leadership (2), pistol

Commisar Krivitsky, pistol

10 x Bolsheviks, rifles

1 x Maxim gun and crew

Armoured car: Armour 1/1/1, 2 crew, 1 x HMG

Weapons
Pistol: range 6 inches, one shot

Rifle: range infinite, one shot

Maxim/heavy machine gun: range infinite, three shots

Special rules
Commissar's discipline (costed in as +3 points): If the Bolshevik army is about to break, the Commissar may temporarily increase motivation by shooting some of his own men at +3pts motivation per man. Remove the 'shot' models from play and count them as casualties the next time the Bolshevik player tests for morale. Note that the armoured car has two crew so this model may be removed to give +6 points motivation but contribute two casualties to subsequent morale checks.

Chapter 5

World War II

'The battlefront disappeared, and with it the illusion that there had ever been a battlefront. For this was no war of occupation, but a war of quick penetration and obliteration
Blitzkrieg, Lightning War.' – *Time Magazine*

Introduction

By 1918, most of the military technology that was to transform warfare was already in place, albeit often immature. Examples include: combat aircraft, radio, tanks, armoured carriers, anti-tank weapons, light automatic weapons, directed artillery fire and so on. The bloodbath of World War I led to the formulation of imaginative but unrealistic theories about how to win a battle by manoeuvre, rather than a sordid war of attrition; the classic example is British cruiser tank warfare.

In practice little in principle had changed. Wars were still won by attrition, destroying an opponent's ability to make war faster than he can destroy yours. Battles were still won by punching holes in, and enveloping part of, or all of, an opponent's defensive line. But radio-controlled, mobile, armoured warfare had shifted the advantage away from the defence and back towards attackers. The Germans were the first to divide their army into mobile divisions and infantry divisions when the rest of the world still thought in terms of cavalry (now in tanks) and infantry/artillery formations.

A pattern developed and spread to all armies where the mobile/armoured divisions attacked and punched holes in enemy defences, whereas the infantry followed up and held ground, enveloping enemy formations. Armoured divisions were all-arms formations with tanks, motorized infantry and towed/self-propelled artillery.

From a skirmish point of view a key change was the move away from the bolt-action rifle as the primary infantry weapon to fighting sections based around a portable light machine gun that could be fired using a bipod or even from the hip. A typical British infantry section might be eight men divided into two groups of four. One brick (four men) would supply covering fire with the light machine gun while the other would skirmish with grenades and rifles.

The German army based its infantry tactics around the Spandau MG34. This was a superb weapon that could function either as a light machine gun or a tripod-mounted heavy machine gun, so it was the world's first General Purpose Machine Gun.

The American army developed a slightly different approach, issuing its infantry with the M1 Garand semi-automatic rifle. These are not different enough from bolt action rifles to warrant different rules. The US Army also used the Browning Automatic Rifle as an LMG during WWII but it was not particularly effective either as a rifle or as a light machine gun. It is treated as a light machine gun in this game because that was how it was used.

Finally the Germany army developed the Sturmgewehr 44 assault rifle during the war when it became clear that most firefights took place at under 400 metres. An assault rifle is switchable between semi-automatic and automatic so it combines the properties of a sub machine gun and rifle in one weapon. The key was to use a lighter cartridge that reduced recoil and allowed more rounds to be carried by an infantryman. This shortened the range but that was irrelevant in practice. The assault rifle has become the standard infantry weapon of the modern world.

The other primary change in skirmish weaponry in WWII was spurred by the tank gun/armour race. Armour got thicker and anti-tank weapons got progressively more powerful until anti-tank units were using monsters like the British 17pdr. The small, man-manoeuvrable light anti-tank guns like the German Pak 36 that had proved so useful in the inter-war years now proved ineffective. The anti-tank rifle quickly became obsolete, although the Red Army continued to use it *en masse* to damage suspensions and cause tank commanders to button down.

The solution was to equip infantry with an entirely new weapon: the rocket-propelled, shaped-charge, anti-tank grenade. That shaped charges could produce directed blast was known from the nineteenth century but the key to an armour-piercing charge was a hollow cone of explosive lined with metal, like a funnel, to produce a focussed plasma stream.

The first use of a shaped-charge anti-tank weapon was the 1940 British no 68 anti-tank grenade, fired from a rifle grenade discharger. It could burn through 50mm of armour compared to the 23mm penetration of the Boyes anti-tank rifle. As tank armour grew thicker, shaped charge grenades became more powerful and rocket or spigot-mortar propelled. Examples include the British PIAT, American bazooka, and German panzerfausts and panzerschreck – the latter could penetrate 160mm of armour at 150 metres. Although range and penetration varied with different marks of these weapons, they are treated as all the same for game purposes.

Various powerful hand-thrown or placed bombs also meant it was very dangerous for a tank unsupported by friendly infantry to approach enemy infantry in cover.

The Soviet Union eschewed shaped charge weapons and continued to develop anti-tank rifles, producing the Simonov that could penetrate 40mm of armour at 100 metres.

Special Rules

Infantry

Assault Rifle: range infinite, one shot OR two shots, range 18 inches.

Simonov anti-tank rifle: range eighteen inches, one shot, armour piercing 2.

Bazooka/PIAT/panzerfaust/panzerschreck: range eighteen inches, one shot, Armour Piercing 4.

Note: Bazooka-type weapons can be used as HE weapons when fired against infantry – minimum range 10 inches, one shot, blast radius 2 inch.

Alternative Armies

The sky is the limit here. Options include: US infantry/airborne/ Marines/ armoured/ Rangers/OSS; British infantry/airborne/Royal Marines/ armoured/Chindits/Long Range Desert Group/SAS/SOE/Home Guard; Soviet infantry/Red Army/navy/NKVD/Partisans; French infantry/cavalry/ Vichy/Free French/Moroccans/Resistance; German infantry/panzer divisions/airborne/SS/Volkssturm/Brandenburgers/Security/'Werewolves'; Italian desert/Eastern Front/Black Brigades; Imperial Japanese Army and Navy marines – and all the forces of the minor allies, Polish, Greek, Dutch, Belgian, Norwegian, Chinese, Finnish, Rumanian and Hungarian.

Scenario 4

Plan Tortue

Introduction

On 1 June, 1944, the French Resistance received a radio message from the BBC, '*l'heure des combats viendra*' or 'the hour of battle comes', followed by '*Les sanglots longs de l'automne*', meaning 'long sobs of autumn', the following day, and '*bercent mon coeur d'une langueur monotone*', 'wound my heart with a monotonous languor', on 5 June. These codes meant that the long-awaited liberation of France was imminent.

The Resistance was tasked with a series of targets marked for destruction: Plan Vert – railways; Plan Rouge – ammunition dumps; Plan Bleu – power lines; Plan Violet – communications, Plan Jaune – command centres, Plan Noir – fuel depots. The final task, Plan Tortue, or Tortoise, was to impede German use of the road network. For this, the Resistance was supported by SOE agents, SAS detachments, OSS groups and Jedburgh teams. The SAS, in particular, were heavily equipped with armed jeeps and, in one exceptional case, a 6pdr anti-tank gun.

The best known Tortue campaign was against the Das Reich Waffen SS Division, partly because the resistance turned a three-day drive from Toulouse to Normandy into a fifteen-day journey and partly because of the atrocities perpetrated by the SS on French civilians in revenge. The Das Reich staggered into Normandy in dribs and drabs and was sacrificed by being fed into Montgomery's mincing machine rather than arriving *en masse* to deal a decisive blow against the British army.

A typical attack would involve the resistance blocking the road with a tree trunk, forcing a German convoy to debus to examine the obstruction before clearing it. Pushing the tree off the road with a vehicle without debussing was too dangerous as the resistance had a nasty habit of hiding mines under the branches. At this point, any Maquis (Resistance fighters) hiding in the woods or hedges along the road would open fire and throw hand grenades, forcing the German infantry to accept combat.

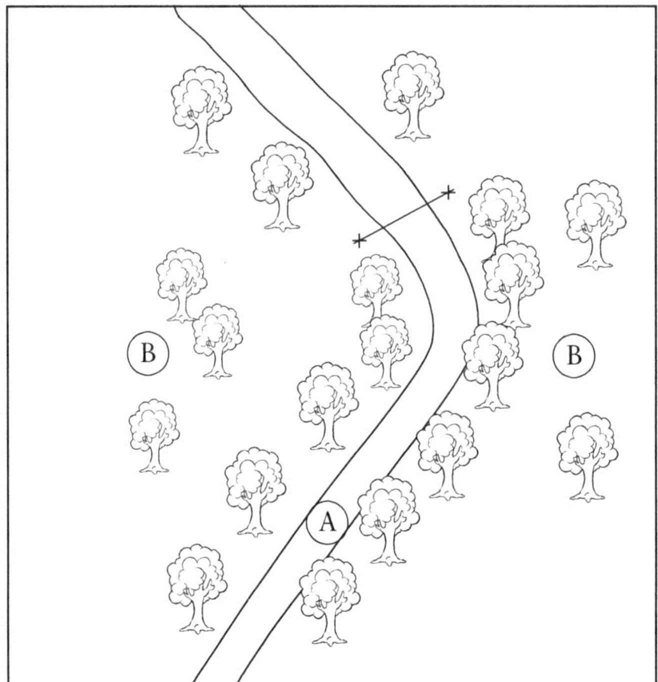

The Battlefield

A winding, European, single-lane road should be laid out across the middle of the map, running south-north. It should be lined with trees and hedges that block off the view along the road at each bend. Rural buildings could also be placed on the map. The road should be obstructed by a tree trunk just north of a bend 12 inches from the southern edge.

The German vehicles start in column on the road south of the obstruction, with their infantry on board. The resistance start the game anywhere on the board more than twelve inches from the road.

Game Length

The game lasts for six turns or until one or both armies break.

Objective
The Germans win by breaking the Resistance army without being broken themselves, or by exiting one vehicle off the north edge of the map by the road. Any other result is a Resistance victory.

Special Rules

Obstruction: The obstruction must be cleared of, or checked for, mines or booby traps before it can be pushed out of the way by a vehicle. This is done by carrying out an action by an infantry model adjacent to the obstruction. It counts as a 'move' even though the model does not actually go anywhere so the player must pay 1, 3 or 5 Action Points as appropriate. The player draws a card and the obstruction is 'cleared' if a picture card (Jack, Queen, or King) is drawn. A single model can have up to three tries to clear the obstruction in a single phase and more than one model may attempt to clear the obstruction in a phase. Once the obstruction is cleared, an adjacent vehicle model can push it out of the way. This counts as a 'move', with the appropriate number of action points being paid, although the vehicle doesn't actually move anywhere.

German Vehicles: These may only move along the road, although the road is wide enough for one vehicle to overtake another.

Resistance Withdrawal: The resistance player may withdraw his models off the east or west side of the board at any time simply by moving them off the playing area. They may not return.

Players Notes

The Resistance player should attempt to protect the obstruction in order to block German vehicle movement as a first strategy to secure a win. The player's second objective should be to destroy the vehicles, rendering defence of the road block irrelevant. If this is achieved, the Resistance player should withdraw from combat immediately to avoid the risk of his army breaking and handing victory to the Germans. However, confident players may eschew such a pusillanimous strategy and go for a double win by destroying the vehicles and then breaking the German army.

The German player should not lose sight of the German objective, which is to get a vehicle off the board, not to kill resistance fighters. However, if the player loses all his vehicles then Himmler's displeasure will only be tempered by breaking the Resistance army.

Forces

Das Reich (52 Points)

Motivation: 4

Vehicles
1 x Armoured Car: Armour 1/1/1, 3 crew, 1 x MG, 1 x Autocannon, Move 24 inches

1 x Hanomag: Armour 1/1/1, 2 crew, 1 x MG, carries 8 passengers, Move 12 inches

Truck: Armour 0/0/0, 1 crew, carries 12 passengers, Move 24 inches

Infantry
SS-Obersturmführer Fluck, Leader (3), Schmeisser

SS-Scharführer Klunkerhaffen, Leader (1), Bruiser (2), Schmeisser, grenades

6 x SS-Schütze (riflemen) with rifles

2 x SS-Schütze (riflemen) with assault rifles and grenades

2 x SS-Schutze, each with 1 MG34 Spandau LMG

Weapons
Pistol: range 6 inches, one shot

Rifle: range infinite, one shot

Assault Rifle: one shot, range infinite; OR two shots, range 18 inches

Schmeisser: one shot, range 18 inches; OR 2 shots, range 6 inches

MG34, Spandau: LMG, range infinite, two shots

Grenade: range 9 inches, one shot, blast radius 2 inches, Armour Piercing 1

Note – only one 'to hit' Resolution Card is drawn when using a grenade. A model can only fire a gun OR throw a grenade once per phase. Vehicle Machine Gun: range infinite, three shots

Autocannon: range infinite, two shots, Armour Piercing 2

French Resistance (52 points)

Motivation 3

SOE Agent
Dame Edith, Leader (3), Inspiring, Sten gun, grenades

SAS
Sgt Benson, Leader (1), Sten gun, grenades

1 x Trooper with rifle, grenades

1 x Trooper with PIAT

Maquis
René, hero of the resistance, Leader (1), Sten gun

Mimi, Tough (1), Sten gun, grenades

M Roger, rifle, Dead Shot (2)

M Alphanck, Scout (1), pistol, grenades

2 x maquis with rifles, grenades

2 x maquis with Sten guns

3 x maquis with rifles

Weapons
Pistol: range 6 inches, one shot

Rifle: range infinite, one shot

Sten Gun: one shot, range 18 inches; OR 2 shots, range 6 inches

Grenade: range 9 inches, one shot, blast radius 2 inches, Armour Piercing 1
Note – only one 'to hit' resolution card is drawn when using a grenade. A model can only fire a gun OR throw the grenade once per phase.

PIAT: range 18 inches, one shot, Armour Piercing 4.

Chapter 6

The Cold War

'Whether you like it or not, history is on our side. We will bury you.'
Khrushchev

Introduction

World War II was probably the last major conflict to be fought with massive industrially sustained and equipped armies. The invention of nuclear weapons and the intercontinental missile made such warfare obsolete. Nuclear superpowers were understandably reluctant to challenge each other militarily because of mutually assured destruction. Instead, superpower rivalry took the form of wars fought entirely by proxies or by one superpower against a rival's proxy state. This is more or less the strategic stalemate predicted by George Orwell's brilliant 1948 novel, *Nineteen Eighty-Four*.

As far as infantry skirmish weapons went, the pattern was for more potent versions of extant weapon systems with higher rates of fire and harder hitting ammunition: eg the modern fully-automatic grenade launcher (AGL) versus the one-shot, rifle-fired grenade launcher of WWII. Armoured warfare continued the pattern set in World War II of bigger guns with more dangerous ammunition opposed by ever thicker and more capable armour.

Tactically, the most innovative change was the development of guided weapons. Nazi Germany had experiments with devices such as the wire-guided Ruhrstahl X4 air-to-air missile and the X7 anti-tank missile. They also experimented with radio-guidance systems such as the Henschell 117 'butterfly' air-to-air and surface-to-air weapons, and the Wasserfall radio-guided surface-to-air weapon. But the only weapons to go into service were the Fritz X and Henschel HS 293 radio-guided anti-ship bombs.

. By 1945, infantry anti-tank weapons capable of piercing the latest armour were either suicidally short-ranged such as the PIAT, or large and clumsy such as the 17pdr. The X7 offered the potential for man-portable anti-tank weapons that could hit and destroy a battle tank at long range. The Soviet Union pioneered the use of portable anti-tank guided missiles as Khrushchev was a missile enthusiast.

First-generation ATGM (anti-tank guided missile) weapons, starting with the French SS-10 in 1955, were 'flown' onto the target by the operator as was intended for the X7 and so were difficult to use. Second-generation weapons such as Milan, introduced in the early 1970s, use a semi-automatic guidance system to 3,000 metres. Modern third-generation missiles, such as Javelin, are fully automated 'fire and forget'.

Special Rules

Infantry

ATGM: range infinite, minimum range 20 inches, one shot, armour piercing 8.

Note: ATGM can be used as Field Guns against infantry, minimum range 20 inches, one shot, blast radius 3 inches.

AGL (automatic grenade launcher): range infinite, three shots, two action cards per shot, crew served.

Note: AGLs operate like machine guns in that the ricochet rules apply for shots but two action cards are turned over for each shot onto a single target – the second card cannot ricochet.

Vehicles

Enhanced Armoured Personal Carrier Armour: 3/2/2

Enhanced Main Battle Tank Armour: 8/4/3

(Enhanced armour refers to modern/late Cold War developments like composite, reactive or electric charge.)

Main Battle Tank Cannon: range infinite, one shot, Armour Piercing 11, blast radius 3"

Alternative Armies

Major regular armies include: NATO, notably the USA, Britain and West Germany; France; the Warsaw Pact, notably the Soviet Union, Poland and East Germany; Israel, Iran, Syria, Iraq and Egypt; and China. Irregular armies include: the Vietcong; various Arab militias; various Afghan militias; various South American revolutionary and counter-revolutionary forces; various Western revolutionary groups and Special Forces; and Chechens.

Scenario 5

Grave Of Empires

Introduction

A Soviet-backed communist government took power in Afghanistan in 1979 in the latest twist of The Great Game and proceeded to purge 'anti-revolutionary' people including army officers. Full-scale civil war broke out. Later that year, the President was assassinated by his Prime Minister triggering a Soviet invasion to 'restore order'. Over nine years of bloody conflict, at least a million Afghans were killed and more than five million forced out as refugees from a population of less than twenty million. The Red Army in the 1970s was equipped and trained to fight major armoured clashes with NATO in the open plains of the Fulda Gap so was fighting at a disadvantage. The Soviet Union itself collapsed in 1991.

In February 1987, Major VA Stolbinskiy was part of a motorized rifle brigade controlling the northwest outskirts of Kandahar City near the border with Pakistan. He was ordered to interdict weapons caravans travelling through *Mujahideen*-controlled territory. Using tight security, Stolbinskiy moved an elite twenty-six-man Recon Platoon, supported by an AGL, into the area while dismounted from their BTR transports. Sixteen men were stationed in the rear to protect the withdrawal route. Five men, including Major VA Stolbinskiy and Sr Lt AV Kholod, mounted an ambush where a culvert from a dry stream bed ran under the road, the other five maintaining overwatch with the AGL and a PK machine gun.

They captured some armed bicyclists without difficulty then attacked a car that was being followed by eight motorcyclists who returned fire. *Mujahideen* began to stream out of Jegdaley in support. The Soviets removed seals and documents from the car before destroying it, and killed five Afghans including Oka, a senior commander, and his adviser, Captain Turan of the Afghan army.

The Red Army group successfully withdrew, one private soldier receiving a shoulder wound.

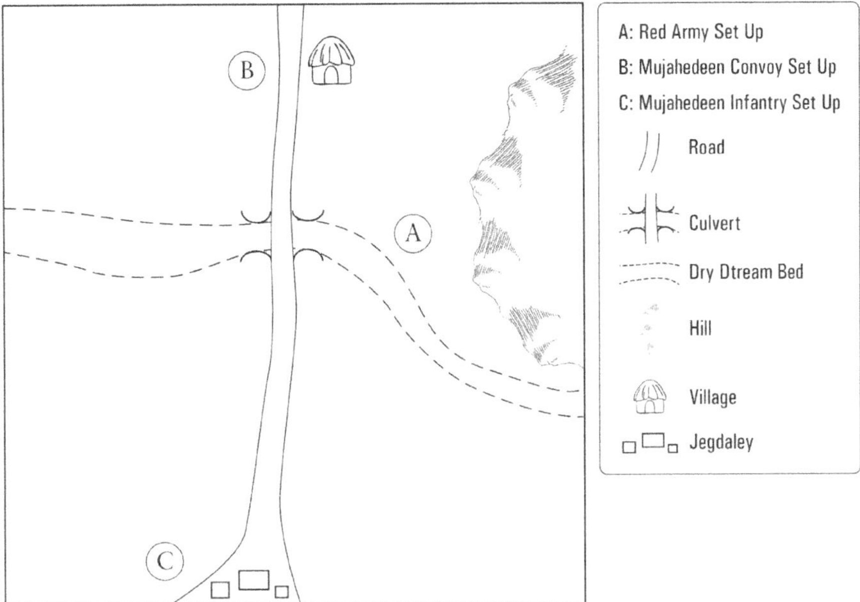

The Battlefield

The terrain is flat and rocky with slight cover apart from on the hill and in the dry stream bed.

The Red Army is placed on the table first at least twelve inches from the village and Jegdaley.

Mujahideen: The car and motorcyclists start on the road, the car in the lead, nine inches north of the culvert. The infantry group are placed in Jegdaley.

Game Length

The game ends when the Red Army breaks or all its surviving troops have exited the table from the east edge of the table, or if both Afghan armies are broken or destroyed.

Objectives

The Soviets receive victory points as follows:

For each Soviet soldier exiting by the east edge   1

Note: if both Afghan armies break and the Soviet Army is unbroken then all remaining Soviet soldiers (including knock downs) are considered to have exited the table.

For capturing the seals and documents   5

For killing Oka   5

For killing Turan   3

The Afghans receive victory points as follows:

For each Soviet soldier killed   2

For each of the two Afghan Armies unbroken at the end of the game   6

Note: 'kill' means kill in close combat or by ranged fire, models removed because their army broke are not 'killed'.

The player who has the highest number of victory points at the end of the game wins, otherwise it is a draw.

Special Rules

Dirt Road: Wheeled vehicles move along the dirt road as if they were 'off-road', i.e. at 6 inches per move and the car can only move along the road.

Ambush: The car and motorbikes can only move along the road at one move per phase until the Soviet player springs the ambush by firing a weapon or engaging in close combat. The *Mujahideen* in Jagdaley cannot move or shoot until the Soviet player springs the ambush.

Seals and Documents: Capturing the seals and documents can only be done by moving one of the Soviet officers adjacent to the car when it is stopped, presumably by the Soviet player wrecking it, and by making a 'move': the model doesn't actually move but searches the car.

Motor Bikes: a model on a motor bike cannot fire a weapon unless it spends the phase stationary but can move adjacent to an enemy model to initiate close combat (riding someone down) drawing an additional card for the weight of the bike. Once a model 'gets off the bike' the model is permanently removed and replaced by an infantry model with the same equipment. A motorbike rider that is knocked down is deemed to have fallen off his machine. Motor bikes are infantry models for the purposes of close combat or being shot at but get no bonus for being in cover. They are wheeled vehicles for the purposes of movement.

Players Notes

Winning this scenario requires each player to keep a close eye on his own and his opponent's victory conditions. The maximum number of points the Soviet player can score is 23 by achieving each of his three objectives and by exiting ALL his troops uninjured. Killing random Afghan warriors and breaking their armies in itself scores nothing. The Afghan player, on the other hand needs to kill Russians. Of course, preventing your opponent scoring victory points is also a useful strategy.

Forces

Soviets (39 points)

Motivation 3

Major Stolbinskiy, PSM pistol, grenades, Leadership (4)

Sr Lt AV Kholod, AK 74, grenades, Leadership (2)

5 x Troopers, AK 74, grenades

1 x Trooper, PK Machine Gun

1 x AGS 17 Team (crew-served weapon with 2 crew)

Weapons
PSM pistol: range 6 inches, one shot

AK 74 Kalash: one shot, range infinite; OR two shots, range 18 inches

PK LMG: range infinite, two shots

AGS 17: range infinite, three shots, two action cards per shot, crew-served

Grenade: range 9 inches, one shot, blast radius 2″, armour piercing 1

Note – only one 'to hit' resolution card is drawn when using a grenade. A model can only fire a gun OR throw a grenade once per phase.

Mujahideen Convoy (33 points)

Motivation 3

1 x Car: Armour 0/0/0, 1 crew, carries 3 passengers, Move 6 inches [see the Dirt Road special rule above]

Commander Oka (starts in the car), AK 47, Leadership (2)

Captain Turan, (starts in the car), AK 47, Leadership (1)
Bodyguard, (starts in the car), AK47, Dead Shot (1)

6 x *Mujahideen*, AK 47, motorbikes

2 x *Mujahideen*, RPG-7, motorbikes

Weapons
AK 47 Kalash: one shot, infinite range; OR two shots, range 18 inches

RPG: range 18 inches, one shot, Armour Piercing 4
Note: can be used as mortar when fired against infantry – minimum range 10 inches, one shot, blast radius 2 inches.

Mujahideen Infantry in Jegdaley (15 pts)

Motivation: 1

Commander, AK47, Leadership (1)

5 x *Mujahideen*, AK47

2 x *Mujahideen*, RPG-7

Weapons
AK 47 Kalash: one shot, infinite range, OR two shots, range 18 inches

RPG: range 18 inches, one shot, Armour Piercing 4

Part III

Extending The Game

Chapter 7

Pulp Action

'Don't move! Or I'll fill you full of … little yellow bolts of light.'

Farscape

Introduction

The Pulp Action chapter covers the worlds of the 'pulp fiction' comics, magazines and novels. Included in this genre are straightforward 'historical' scenarios featuring gangsters, explorers, gentleman detectives and the like, which we have essentially covered in earlier chapters. So here I want to focus on what, for the sake of a label, we might call the 'Astounding Stories' end of the market. Here, square-jawed heroes and beautiful heroines battle evil masterminds who have at their disposal magical or weird-technology, in so far as they can be distinguished, and a retinue of heavies. We are in the company of superheroes, doom temples, moon Nazis, cultists and mad professors.

The use of magic/weird technology in skirmish games breaks down into four principles: weapons, mind-controllers, open sesames, and summoning unpleasantness.

Magical weapons are perhaps the easiest and most boring to include in one's skirmish games. A lightning bolt, psychic hit, or magical arrow are not terribly exciting if they behave just like a rifle only with increased probability of a knock-down, so try to give them some new quality. Mind controllers offer more potential in that they can work quite differently from a physical attack, bypassing the target's normal defences.

Open sesames are unusual abilities that give a character unique access to something. Examples might include: a sorcerer who knows just the right spell to unlock the magic door to the treasure chamber under the temple of doom; a time-traveller with a sonic can-opener who has both the knowledge and the appropriate gizmo to reverse the polarity on the doomsday machine that is about to destroy the galaxy; or a mad professor whose knowledge of archaic, lost languages allows him to decipher the clay tablets so as to work out where the mad cultists are seeking the Hand of Thoth – possession of which will doom…; well, I'm sure you get the idea.

Doom usually features rather strongly. The open sesame ability makes the specific character uniquely important to the scenario, which will have a major impact on the tactics employed by both players. Indeed, loss of this character could well mean loss of the whole game. I would recommend giving such a character the 'Lucky' special capability.

Finally we have: summon various sorts of unpleasantness. This might be an SS *Ahnenerbe* sorcerer summoning an earth-elemental using arcane Aryan knowledge, or a mad professor of geology with an invention of his own devising that causes earthquakes. In game terms the results might be rather similar, for example all models within twelve inches having to pass an action card test against a fixed number or be knocked down.

Special Rules

The main issue with putting these various abilities and gadgets into our games is making them powerful enough to matter but not so powerful that they utterly dominate the scenario. Then working out how many points they should cost. There are three variables under our control as a scenario designer: the ubiquity of the effect, the strength of the effect, and how often the ability can be used.

Considering ubiquity, if the effect spreads out in all directions over a large area, say a circle of radius 12 inches around the model targeting anything in its path, then that will have much more effect than an ability that can only target one model with a range of 6 inches. Similarly an ability that draws five action cards to hit is more powerful than an ability that draws one. The three choices for how often an ability can be used is every phase, every turn, or once per game. Clearly the first selection makes the ability far more powerful than the last.

Alternative Armies

The universe is your oyster when it comes to armies for pulp scenarios. You have at your disposal every pulp story every printed, or shown on TV or at the cinema. My favourites include: retro 1950s sci-fi, Victorian SF, Steampunk, pre-WWII Nazi archaeological weirdness, post WWII Nazi strange-science weirdness with flying saucers and zero-point energy bells, religious cults summoning Satan, ancient Gods, daemons etc, Lovecraftian SF, hollow-Earth explorations and other dinosaurs, and, and, …… oh, too many to list.

Scenario 6

Retro Rockets

Introduction

The Teacon, Emperor of Venus and Commander of the feared Troon shock troops, has kidnapped Professor Slimbody, brilliant science advisor to the Royal Air Force, in an effort to force her to reveal the secret formula used in the special brand of tea served in RAF messes. Dick Dangerous, pilot of the unlikely future and his batman Wigsby, with a squad of Royal Marines, blast off from their Lancashire spaceport to rescue the Professor.

Unfortunately, Wigsby told his hairdresser about the mission just before take-off not knowing that the coiffeur was a Troon spy. Upon hearing the news that his hated nemesis was on his way to Venus, the Teacon arranged for Professor Slimbody to be held in a cage in the middle of the dinosaur marsh as bait. The evil mastermind arranged his Troon soldiers out of sight near the cage.

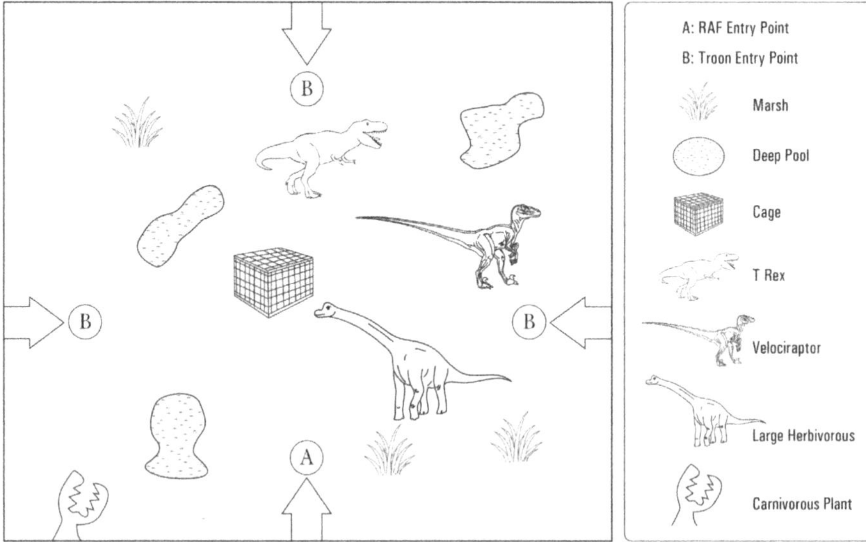

Dangerous landed his Spaceship, *Nikolayevna*, on some firm ground on the edge of the swamp, having located the beacon hidden about the Professor's person. Dick reasoned that not even the evil Teacon would be such a bounder as to search a lady. But little did our hero know that the mastermind had known about the beacon all along; indeed, he was counting on it. What the Teacon didn't know is that Slimbody also had a needle pistol concealed about her person, a tiny but powerful weapon of her own invention.

Can our gallant lads rescue Slimbody and fight their way out of the deadly trap? Stay tuned to this channel!

The Battlefield

Set up the cage in the middle of the playing area. There are no playing models on the table at the start of the game. The playing area should have patches of marsh and the odd deep-water pool that is uncrossable. It should be rich in vegetation. Scatter a few carnivorous plants and the odd dinosaur around at random to add spice, but they must be more than 6 inches away from the cage.

RAF models enter from the south edge of the table, while Troon models may enter from the other three edges at least 6 inches from the south edge, by making a normal move. RAF models may leave by the southern edge. Troon models may not leave the playing area.

Game Length

The game lasts for eight turns, or until Slimbody has left the playing area by an edge, or an army breaks.

Objective

A player wins a sudden death victory if they break the opposing army while their own army remains unbroken. Otherwise, whoever has scored the most victory points when the game ends is the victor. Otherwise, it's a draw. Victory points are awarded as follows:

RAF

Freeing Slimbody from the cage	3pts
Slimbody walks off the South table edge	5pts
Slimbody carried insensible off the South table edge	3pt
Killing the Teacon	3pts

Troon

Slimbody still in cage at the end of the game	5pts
Slimbody insensible and on the table at the end of the game	3pts
Carrying Slimbody insensible off the West, North or East table edge	4pts
Killing Dick Dangerous	5pts
Killing Wigsby	2pts

Special Rules

Weapons:

Ray Guns: range 18 inches, one shot, Armour Piercing 3. Ray guns are peculiar things in that their range and power are entirely the same irrespective of the size of the weapon or whether it is configured like a rifle, pistol or hoplite spear (it does happen, the Atlantean Guard for example). They do tend to hit hard so any model downed by a ray gun is immediately removed from the table, rather than knocked down, if the difference between the attacking card and the highest defensive card is equal or greater than +5.

Laser rifle: range infinite, two shots, Armour Piercing 1. Laser rifles are not much more powerful than an ordinary rifle but, as they fire a line of energy, they may be 'walked' onto the target giving them the equivalent of two shots.

Needle Pistol: range 6 inches, one shot. The pistol fires tiny darts that inject a powerful nerve toxin. The target is not knocked down if it fails to successfully defend against a shot, but is poisoned and permanently removed from the game.

Plasma Bazooka: range 20 inches, one shot, Armour Piercing 8.

The Teacon

The evil one rides around up to 6 inches per move on a flying anti-gravity disc, which he controls by telepathy. Note that because the disc flies, it is not impeded by terrain. The disk is equipped with a number of concealed weapons, only one of which may be used per phase at a cost of one action point.

Ray Gun: the disc has a built in ray gun (as above)

Stunner: the disk can transmit a bubble of energy that attacks the brain of all non-vehicle models within 'X' inches using one action card; draw an action card to find out what 'X' is. This is a shoot action. The attacked models defend with one action card each with no additional cards for terrain. Note that this weapon can only stun so all affected models automatically recover at the end of the turn and are stood upright.

Psychic Amplifier: range infinite, three shots, no ricochets. The disc has a psychic amplifier that can boost the Teacon's already-considerable mental powers. He may target a single model at any range within his line of sight and attack with three cards. This is a shoot action. The model gets one defensive card, irrespective of the terrain, or three defensive cards if it is a named character (such as Wigsby or Dick Dangerous). If the Teacon wins, he gains control of the enemy model momentarily and may force it to make a move, including into close combat, or a fire action.

Teacon Points: it is probably worth considering how the Teacon points are arrived at as this is quite a complex character and showing how I arrived at the final points total may be helpful to readers creating their own characters. The basic model upgraded with a ray gun is 2 pts. Making him a Leader (2) costs a further +4 pts. The anti-grav disc is a useful device, particularly in a marshy environment, so I value it at +2 pts. The stunner is potentially a devastating weapon but it is unreliable (very variable range) and is not in itself lethal so I value it at +2pts. Similarly the psychic amplifier could be nasty in the right circumstances but can only affect one target so I cost it at +1pt. As a general note, because these three weapons all require a shoot action, the Teacon can only use one per phase, to some degree limiting their usefulness. So we end up with an, um, eleven point character. I am rubbish at arithmetic – if God meant us to be able to count, He wouldn't have given us computers.

Other Stuff

The Cage: Any RAF model that is adjacent to the cage may attempt to unlock it once per phase by paying an action point and testing with one action card. The model must beat a score of 5. Dick Dangerous gets a +3 bonus. The cage is energy shielded so it de-powers when unlocked. Remove it and replace with a model of Professor Slimbody that may be activated like any other RAF model.

Professor Slimbody: Professor Slimbody may not be shot at by the Troon Army as the Teacon has not yet obtained the secret formula so he needs her alive. So she cannot be killed in this scenario. If she escapes, she may be recaptured by a Troon model by being defeated in close combat, but note that Slimbody is a judo expert! She is then insensible, and can only be moved by being 'carried' by a Troon model. She is simply moved along with the Troon model as it moves. She may be re-recaptured by the Troon being defeated in close combat and killed by an RAF model. She can then be 'carried' by the RAF model.....and so on.

Here Be Monsters

Carnivorous Plants: A player may try to activate one carnivorous plant per player-phase by paying an action point. Draw a card: if it's a Jack the player may attack one enemy model within 2 inches range of the plant; if a Queen, 4 inches; if a King, 6 inches. Other cards mean that the plant fails to activate. To attack, the plant sends out tendrils that pull the model adjacent, then makes a standard close combat attack.

Dinosaurs: Dinosaurs are classified as large carnivores such as a T. rex, small carnivores such as a velociraptor, or large dangerous herbivores such as a triceratops; any other type of dinosaur is simply immobile terrain. A player may attempt to get control of one dinosaur per player-phase by paying an action point. Draw a card: a small dinosaur is activated by beating '5', a large carnivore by beating '8' and a large herbivore by beating a '10'. If successful, the player may move the dinosaur once, paying such terrain modifications costs as apply. Small dinosaurs move 10 inches, large carnivorous dinosaurs 6 inches and large herbivorous dinosaurs 4 inches. A small carnivore makes a close combat attack drawing one card, a large carnivore three cards, and a large herbivore five cards. If the dinosaur loses the fight it is removed from the table as usual. Treat large dinosaurs as vehicles for the purposes of being shot at, with armour 2 all round.

Marsh: models move across marsh areas at half speed, ie it takes 2 inches of move to move across 1 inch of marsh – or part thereof.

Players Notes

Clearly, the game hinges on what happens to Professor Slimbody but both players should pay particular attention to the victory points. Ideally the Troon player wants to keep Slimbody in the cage because the RAF player can then only win by breaking the Troon army. Similarly the RAF player wants Slimbody to walk off the table under her own steam for a decisive victory: three points for freeing her from the cage plus five points for the Walk Off.

But if the RAF player carries Slimbody off the table for six points (3+3), then the Troon player can still win by killing Dangerous and Wigsby for seven points – unless the RAF player kills the Teacon as well. If neither player manages to get Slimbody off the table then subsidiary objectives can make the difference. It will come down to who dies and whether Slimbody is insensible when the game ends.

Incidentally, gamers 'of a certain age' may notice that this is a simplified parody. Suffice it to say that I was a keen reader of *The Eagle* comic in the late 50s and early 60s. Those who want to recreate more realistic scenarios involving Dan Dare, pilot of the future, with all the richness of the original, are recommended to look up reprints of the comic strip available from book shops or get hold of a copy of the Haynes Space Fleet Operations Manual.

Forces

RAF (42 Points)

Motivation: 3

Dick Dangerous, Leader (5), ray gun, Lucky

Professor Slimbody, needle pistol, Judo Expert

Wigsby, pistol

Sgt Smith, Leader (1), laser rifle

8 x troopers with laser rifles

One trooper with plasma bazooka

Weapons
Pistol: range 6 inches, one shot

Needle pistol: range six inches, one shot.
Note: The target is poisoned and removed if knocked down

Ray gun: range 18 inches, one shot, Armour Piercing 3

Laser rifle: range infinite, two shots, Armour Piercing 1, ricochet

Plasma Bazooka: range 20 inches, one shot, Armour Piercing 8

Special Rules
Lucky: knocked down model only lost on the drawing of a Heart card, rather than any red card

Judo Expert: draws two action cards for defence when attacked in close combat

Troon (43 Points)

Motivation: 7

Teacon: Leader (2), anti-grav disc, ray gun, stunner, psychic amplifier

Troon Commander: Leader (1), ray gun

9 x Troon troopers with ray guns

Weapons

Ray gun: range 18 inches, one shot, armour piercing 3

Stunner: range (Action Card) inches, one shot every model within range, no additional defence cards for terrain, weapon only stuns so affected models recover at the end of the turn and are stood upright.

Psychic Amplifier: range infinite, three shots, no ricochets, no additional defence cards for terrain.

Chapter 8

Fighting a Campaign

'In war, then, let your great object be victory, not lengthy campaigns.'

Sun Tzu

Introduction

Before getting down to detail, it is worth making a few general remarks. Campaigns are things that are often started in many wargaming clubs but rarely finished. No campaign that requires getting a half dozen people or more to play a set game each week is likely to last out the month. The problem is that people have commitments; even wargamers have something resembling a life. They have work commitments, family commitments or just get bored and pursue some new gaming fad. In the same spirit, the likelihood of a campaign being abandoned increases exponentially in proportion to its complexity and any campaign, no matter how brilliant, will simply peter out if it has no fixed end point. It is important to have a conclusion where someone is adjudicated the winner.

It is a different matter when the campaign is played between two long-term friends who are accustomed to meeting once a week, or whatever, to play a game. Such projects can be as simple or as complex as the players' desire and may last as long as they choose. Even finding a winner might not matter too much. But matters become a little more complicated, when three or more are gathered together to fight.

Victory in a club campaign should be marked by some sort of Triumph for the winner so as to incentivize the players. It is, in my opinion, a mistake to make this something of more than nominal financial value. The award could simply be an announcement and round of applause or something tangible like an old, battered but gold-painted plastic space marine, or maybe the right to wear a victor's cap (and be jeered) at club meetings for the rest of the year.

Ladder Campaigns

The simplest form of campaign involves club members playing as many or as few games as they like, in no particular order, and against whichever opponent is conveniently available on the day. Also, people may play whatever scenarios they fancy in any order that they find congenial and as many times as they like. One might award players a point for a win and, at some specific date such as the club annual general meeting or the last meeting of the year, tot up who has the most points to announce a *Victor Ludorum*.

The problem with this approach is that it favours the player who plays the most games. Playing fifty games but only winning five would give a player a higher score than someone who plays just four, nevertheless winning them all. Of course, one can always simply convert wins into a percentage by dividing wins by games played and multiplying by one hundred but even this is subject to mathematical gamesmanship: a player who wins his first game might be well advised to play no more as he has already achieved a perfect, unbeatable, one hundred percent score.

One fun way of getting around such imperfection is to set up a club ladder. One starts by writing the names of all participants on cards, shuffling them, and pinning them to a board in a line. If a participant beats a player whose card has a higher position up the ladder then the two cards change places. So if the bottom player on the ladder beats the person at the top they immediately become *numero uno* and visa versa. The better players will tend to rise up the ladder and stay there but, but… there is always the chance of an upset giving spice to the proceedings. Oh and any player who doesn't play a game within a set time, say a month, drops a ranking, exchanging card positions with the person immediately below them. Obviously, a ladder campaign should end on a set date to add some drama to the proceedings.

A variant on the ladder campaign is to group players into 'houses' or teams of mixed ability and experience. The team, not the individual is represented by a card on the ladder. For team games, it is best to dampen down results by having a winner who plays a higher ranked team exchanging card places with the team immediately above them, even if it wasn't the team they were actually playing.

Map Campaigns

A common and popular variant of the wargaming campaign is to use maps upon which armies move from point to point, or area to area, attempting to win the game by conquering and acquiring territory. Obviously such conquests are contested by other players leading to battles. The area/nodes on the map should have different terrain that can be replicated on the tabletop.

We must never lose sight of the fact that the primary *raison d'être* of a campaign is simply to provide an interesting framework for tabletop battles. The campaign must never become so complicated that it becomes a game in its own right, detracting from wargames.

Map campaigns are not usually considered to lend themselves to skirmish wargaming because of the small geographic scale covered by each encounter. This is because the places covered by skirmish scenarios in a campaign are commonly disconnected rather than contiguous. So it's unlikely that a squad from each side would move across a map bumping into each other in the same way that two vast armies might.

However, it is not impossible to come up with a rationale for a map-based skirmish campaign if we use a little ingenuity.

Map Campaign – Escaping An Encirclement

Introduction

The year is 1943 and in the East, German forces are in full retreat after their failure to break through Soviet defences around Kursk. On 12 July, the Red Army launched Operation Kutusov on the north face of the Kursk salient. Two spearheads smashed through the German lines north and south of Orel. A small unit of German soldiers has been cut off by one of these spearheads and is attempting to escape westwards through Soviet blocking forces.

The Germans must cross to the west of the railway line running north-south but can't just march down the east-west road as they have to avoid an impassable Soviet road block set up east of the stream. Of course, the Red Army also has patrols out looking for German stragglers.

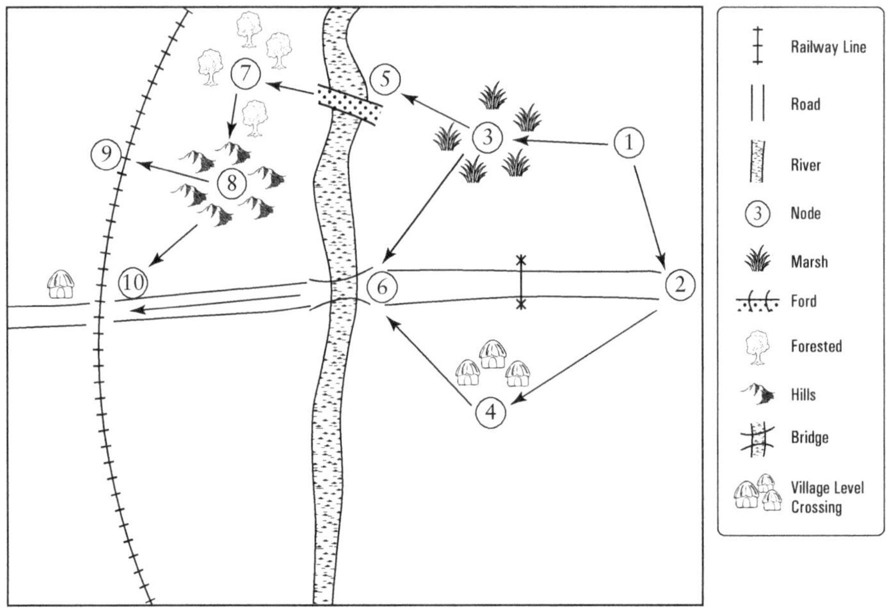

The Battlefield

The map shows two potential German escape routes to the west from their starting position on Node 1 that bypass the Soviet roadblock. The northern route is the slower, crossing the stream at a ford. This path requires moving through Node 3 marsh, Node 5 ford, Node 7 woods, Node 8 hills and Node 9, the railway embankment. The southern route goes across the road at Node 2, through the village at Node 4, across the river at the road bridge at Node 6, and over the railway crossing at Node 10, where there is a signal keeper's house and a few peasant huts.

Each campaign turn this army moves one node along the marked arrows, triggering a tactical battle if it encounters a Soviet patrol in that node. Note that at Node 3 the German player can move to Node 6, a short cut, and at Node 8 can move to Node 10, both transferring his forces from the northern route to the south. The shortest route is 1, 3, 6, 10 – but the Soviet player knows that as well.

Tactical Battles

Each tactical battle involves the Germans setting up on the east side of the table and any Soviet force setting up on the west side. The battle ends when (i) one or both armies break, (ii) all German models not dead or currently knocked down have exited from the west side of the table or (iii) after six turns.

Forces

The German player chooses a single force that he uses for the entire campaign. I would suggest somewhere between 25 and 50 points. At the end of a tactical battle, all the German models that are upright or knocked down on the table, or that have exited the west side of the map, are considered to have survived.

German models that have been 'killed' during a tactical battle may have just run away or hidden – or they may really be dead or seriously wounded. Re-test for each model 'killed' during the battle. On a black card it has survived and re-joins the German force; on a red card it really is dead/seriously wounded/captured and is permanently lost from the German army. The escaping German force will therefore tend to diminish from battle to battle.

Soviet forces are handled somewhat differently. The Soviet player creates four 'patrols', each of which is about 2/3 the points of the German force – so If the German force is 42 points then each Soviet patrol will be 28 points – and two patrols of the same size as the German force. Each Soviet patrol should be as different as the player's models permit to add interest to each battle. The Soviet player allocates each patrol to one of the nodes on the map – not Node 1 – leaving three empty. Soviet patrols never move. They have their orders and the NKVD are ever ready to make sure they obey without question.

Winning And Losing

The campaign ends when all the German models are dead, a battle has been fought in Node 9 or Node 10, or a single German model has exited the west side of the table at Node 9 or Node 10.

The Soviet player wins a strategic victory if he kills all the German models. The German player wins a strategic victory when a German model exits the west side of the table at Node 9 or Node 10. Any other result is a strategic draw.

The German player wins any tactical battle if he exits at least one model from the west side of the board or breaks the Soviet patrol. The Soviet player wins a tactical battle if he breaks the German force or none of the German models have exited the west side of the table. Players get a tactical point for each tactical battle won. Whosoever has the most wins a tactical victory, irrespective of who wins a strategic victory.

Narrative Campaigns

Introduction

A narrative campaign is a set of battle scenarios that are set in a linear story. Sticking with our Eastern Front theme, we might choose to replay a couple of weeks in the life of a German and Soviet squad in the battle for Kursk, 1943. Kursk is commonly depicted as a tank battle but actually infantry predominated, like all WWII battles.

Each encounter is pre-written and whoever wins or loses has no impact on the next encounter. Players may choose different armies for each scenario. The games follow a story as is illustrated below.

Scenario A: Intelligence Gathering

Before the main assault it was normal for an army on the Eastern Front to raid enemy lines in order to gather intelligence about their opponent's dispositions. At its simplest, this might involve a scouting operation to discover the location of prepared defences such as trench lines or bunkers but the ideal was to capture documents or a 'tongue'.

A 'tongue' is Red Army slang for a prisoner who, after being carried back to friendly lines, might be 'persuaded' to give information. Even a private soldier might know a surprising amount of useful information such as: which unit he is from, what units are on each flank, what defences are prepared and their location, general strength in his zone, morale of the defenders, and so on.

In this scenario, a German reconnaissance team is sneaking forward when something goes wrong and the Soviet defenders are alerted. Nevertheless the team press on to complete their mission.

Place two objective markers along a north-south line in the middle of the table, equidistant from the table edge and each other to represent observation points. In the middle of the eastern (Soviet) half of the playing area, place a documents objective marker in a command post.

The Soviet army sets up anywhere in the eastern half of the board. The German player sets up anywhere within 3 inches of the western table edge. The game is player for six turns or until an army breaks. Victory is decided by how well the German player has carried out the reconnaissance, i.e. how many points he has accumulated.

During play, the German player receives 1 point for moving a model to touch an observation point marker: the marker is then removed from play. At the end of five turns, the German player receives 5 points if he has the documents and 3 points for capturing a prisoner (more than one prisoner has no effect on the points received). He receives a bonus of 3 points for carrying a prisoner off the western table edge and a further bonus of 5 points for taking off the documents.

The Soviet player receives 7 points for breaking the German army, 1 point for each German model killed and 3 points for killing a leader.

Whoever has the most points at the end of the game wins.

Scenario B: German Assault

This scenario recreates a tiny part of an initial German offensive against a Soviet prepared line.

Set up the Soviet Army in light improved positions, such as trenches, anywhere in the eastern half of the table. The German Army starts the game within 3 inches of the western table edge. Place five objective markers on the table reasonably spaced out but away from the table edges: three should be in no-man's land between the armies and two within the Soviet half.

If neither army has broken before the end of turn six, deciding victory, then whichever side holds the most objectives wins. An objective is held by the army which has the closest infantry model to it provided the model is not more than 3 inches away.

Both armies should cost roughly the same number of points. Given that the Germans are attacking, then this is an opportunity to use an armoured fighting vehicle. However, if this option is taken, the Soviet player should select anti-armour weapons as part of his army.

Scenario C: Pursuit

The Germans have broken through the line and are pursuing the defeated Soviet units that are in full retreat towards the next prepared position somewhere to the east of the battle area. Obviously both armies may only field vehicles or models transported by vehicles.

The Soviet Army sets up anywhere in the western half of the table. The German army starts the game off table and enter the game using a normal move. On turn 1, German models may only enter from the western table edge. On turn 2 German models still off table may enter by the western table edge, or along the northern or southern table edges 6 inches from the western edge. On turn 3 German models still off table may enter by the western table edge, or along the northern or southern table edges 12 inches from the western edge. Any German models not on the table by the end of turn three are considered to be dead – with the resulting impact on army break tests. Soviet models may leave the playing area by the eastern table edge.

The game ends after six turns or when an army breaks. At this point assess victory by comparing how many Soviet models have left the eastern table edge compared to how many soviet models have been killed. The highest number wins.

Scenario D: Ambush

Eventually, the German assault runs out of momentum as it encounters serious Soviet opposition, ambushing the German spearhead. The German player sets up first and must put models within a 12-inch diameter circle in the centre of the battle field. The Soviet player may place models anywhere outside of the German setup circle provided they are at least 6 inches from a German model. This scenario is fought until one or both armies break, deciding the winner. All German models must be vehicles or models transported by vehicles.

Soviet Counterattack

Now play the same four scenarios again with the roles reversed. Award 1 point for each scenario won and the player with the most points at the end of the game wins the campaign.

Complex Campaigns

Introduction

Skirmish game campaigns lend themselves to a more complex campaign system where units have a history, ie the events of previous scenarios influence the composition of the force in future scenarios. Some models are permanently killed, replacements arrive, and survivors gain experience. So an army changes through time, from scenario to scenario. This type of campaign benefits from being run by a neutral referee or 'games master', who may or may not participate in the battles, if more than two players are involved.

Setting Up An Army

The first thing to do is to set up the initial army with which each participant will play their first battle. It is probably best to start with a fairly small force of perhaps 15 points – as armies will undoubtedly expand as the campaign progresses. The army should only have one leader who may not be higher than Leader (2) and no model can be given Special Capabilities. Other than that, players are free to choose their own mix. I suggest all armies should start with Motivation 1 or 2, which must be paid for from the 15 points.

End Of Battle Mechanisms

At the end of a scenario, a player's army is withdrawn for rest and refit. Place all the models that have been killed to one side. Readers will recall that 'killed' in a scenario just means 'gone missing' in this game. Re-test the 'killed' figures to see if they turn up after the battle. On the turn of a black card they re-appear in a player's army but are permanently lost from the campaign on the turn of a red card: they are dead, too badly wounded to continue, been captured by the enemy, or deserted.

Next, replacements come up the line on a one-for-one basis. Normal trooper replacements have no special capabilities and are armed only with a basic weapon: ie a musket, rifle, assault rifle or ray gun, according to the era. Leader replacements are always Leader (1), have no special capabilities, and have an appropriate basic weapon.

Next work out how many experience points the army has gained from the scenario. These depend on the points-level of the opponent, compared to one's own force, as well as whether the army won, lost or drew the scenario. The logic behind this is that an army learns more by fighting a significantly more powerful opponent than a weaker opponent. As an aside, I would say the same logic applies to wargamers. One learns far more by playing stronger opponents than ones of similar or lesser ability. The other reason for using a graded experience point system is that it tends to handicap the best – or luckiest – players, keeping the interest in the campaign of less successful players. Golf uses a similar handicap system for more or less the same reason.

Experience Points

	Win	Draw	Loose
Opponent 5 or more points higher	7	5	3
Opponent of similar points value	5	3	1
Opponent 5 or more points lower	3	2	1

These experience points can be expended immediately or 'saved up' from scenario to scenario to 'buy' an expensive item. They can be expended on additional troopers or leaders, or to upgrade weapons or capabilities of additional or current models.

Winning And Losing

Armies will get steadily larger, better armed, and more skilful as the campaign progresses. The first player to upgrade his army until it is worth, say, 30 points wins the campaign. You don't want to let it run on until one player has an army so unbeatable that no one wants to play them.

If at the end of a run of scenarios no player has reached the 'sudden death' number of points then award victory to the player who has won most scenarios, using army points value as a tiebreaker.

Chapter 9

Points System

'People know the price of everything and the value of nothing.'

Oscar Wilde

Introduction

The points cost of various models and weapons are one of the most contentious issues in wargaming. The truth is that any points system doesn't work very well because of the wide flexibility and variation in miniature wargaming. The 'true value' of any particular model and weapon will vary quite considerably depending on a whole range of circumstances, notably: the scenario being played and the objectives for each player to win; the terrain, both type and amount; your and your opponent's army composition; your and your opponent's playing styles; size of playing area; and so on, and so on.

This is why the relative 'cost' of different models is such an endless source of disagreement between players. Wargame designers are not ignorant of the issue and, aware that they are on a hiding to nothing, rules writers periodically attempt to drop the whole idea; the most recent example coming from no lesser company than Games Workshop. Generally, such initiatives are welcomed by wargamers like a dose of bubonic plague and points are hurriedly shoehorned back in by chastened publishers.

The simple reality is that players *need* points to create their own scenarios. Any unfairness and imbalance in the resulting games can happily be blamed on the designer rather than by accusations that one's opponent is 'cheating'. Thus harmony, or at least a reasonable facsimile of same, is restored within the wargaming community when someone's master plan goes horribly wrong. The alternative, that one just played badly, is obviously not to be countenanced.

Actually, skirmish games are less subject to design balance problems than most other types because the high level of instability inherent in skirmish mechanics, makes balance kinda difficult to assess by the players as well as the game designer. The system used here is particularly robust to army-balance issues because both the command control and the morale breakdown of an army is geared to absolute numbers of models lost rather than a percentage – so having vast forces hiding in the woods isn't much of an advantage.

In that spirit, I list a points cost for most of the various weapons that players will commonly use – and the rest can be interpolated.

Model Points Cost

Army Motivation per point of motivation	+ 1pt
Model lacking firearm:	½pt
Model with one-shot pistol or rifle:	1pt
Upgrade rifle to anti-tank rifle	+ 1pt
Upgrade rifle to Simov anti-tank rifle	+ 2pts
Upgrade rifle to bazooka, etc.	+ 3pts
Upgrade rifle to ray gun	+ 1pt
Upgrade rifle to laser rifle	+ 1pt
Upgrade to needle pistol	+ 1pt
Upgrade rifle to plasma bazooka	+ 4pts
For every additional shot from a multi-shot infantry weapon:	+ 1pt
Flamethrower	+ 3pts
Cavalryman with sword:	1pt
Cavalryman with lance:	2pts
Cavalryman with one-shot firearm:	2pts
Napoleonic cannon with crew:	2pts
Hand cranked multi-shot weapon with crew:	2pts
To equip with hand grenades:	+ 1pt
Tripod-mounted heavy machine gun and crew:	3pts
AGL with crew	5pts
ATGM with crew	5pts
Mortar with crew:	3pts
Field Gun with crew:	4pts
Motor Bike/bicycle:	+ 1pt

Unarmoured vehicle with crew:	2pts
Armoured carrier/car with crew:	4pts
Light tank with crew:	5pts
Medium tank with crew:	6pts
Heavy tank with crew:	7pts
Main Battle Tank with crew	9pts
Enhanced Armoured Personal Carrier Armour	+ 1pt
Enhanced Main Battle Tank Armour	+ 3pts

Cost to add weapons to vehicles

Light machine gun:	+ 1pt
Heavy machine gun:	+ 2pts
Automatic cannon, per barrel:	+ 3pts
Light tank gun:	+ 3pts
Medium tank gun:	+ 4pts
Heavy tank gun:	+ 5pts
Tank killer:	+ 6pts
Main battle tank gun	+ 8pts

Special Capabilities

Leader (X):	+ (2X)pts
Dead Shot (X):	+ (X)pts
Bruiser (X):	+ (X)pts
Tough (X):	+ (X)pts
Scout (X):	+ (X)pts
Inspiring:	+ 1pts
Fast:	+ 1pt

Lucky: + half the model's initial points, rounding fractions down

Vehicle Special Capabilities

Heroic Commander (X):	+ (2X)pts
Ace Gunner (X):	+ (X)pts
Veteran Driver:	+ 1pt

Army Special Capabilities

Motivation (X):	+ (X)pts
Scenario-Specific Special Capability (X):	+ (X)pts

Examples

Some examples are given below to demonstrate the system:
- British Corporal, Bruiser (1), with musket: 2pts (1+1)
- French Hussar with sabre, carbine and pistol: 2pts
- Zulu induna with spear and shield, Leader (2), Tough (1): 5½pts (½+4+1)
- WWI Tommy with Lewis gun: 2pts (1+1)
- Russian Civil War Putilov-Garford armoured truck/car, 3 light MGs, light tank gun: 10pts (4+3+3)
- Volkswagen Kübelwagen, light MG: 3pts (2+1)
- Michael Wittmann's Tiger tank, Heroic Commander (3), 2 light MGs , Tank Killer: 21pts (7+6+2+6)
- Sgt Gordon's Sherman Firefly, Joe Ekins Dead Shot (3), 1 light MG, Tank Killer: 19pts (6+6+1+6)

Chapter 10

Additional Rules

'There are no absolute rules of conduct, either in peace or war. Everything depends on circumstances.'

Trotsky

Introduction

If you've worked through the basic game from the simple Napoleonic to the complex sci-fi pulp fiction scenario then this chapter introduces some new concepts to put a novel spin on your battles. All of them work to constrain players, in one way or another. The three areas I would like to touch upon are visibility, random events and deteriorating environments

Visibility

So far, our battles have taken place entirely in daylight in good visibility, which is why a rifle at this game scale has a theoretical infinite range. In practice a World War II rifleman had a lesser range than a Zulu War rifleman because in most circumstances he would be lying down while firing rather than standing erect – which is why Germany was more impressed by rate of fire than theoretical effective range when designing the first mass-produced assault rifle.

However, it is far from unknown for battles to occur in minimal visibility, most commonly because they happen at night, but inclement weather such as fog, heavy rain or snow can also seriously impede visibility. Battle itself can adversely affect range of vision by smoke, black powder guns or burning vehicles, trees and cities, or by the dust raised by shellfire, moving vehicles or horses. With a full moon and a clear sky, the ground at night can be surprisingly well lit up but the same landscape under ten-tenths cloud is much gloomier.

Technology has occasionally been used in an attempt to circumvent some of the impact of smoke or night on vision. Various attempts were made to illuminate a night battlefield in World War II, for example search lights to assist a British tank attack in Normandy. Front line armies in the Cold War routinely used infra-red searchlights and receivers on vehicles, adopting technology invented at the end of World War II by German engineers. None of these methods were particularly effective until the advent of modern electronics and thermal sights.

The simplest way to introduce visibility limitations into a game is for both players to agree on a maximum visibility range before the game starts. For example, players might agree that a raid to obtain intelligence on the Eastern Front in WWII is taking place at night and that no model can see further than 12 inches. An element of randomness can be introduced into this by the draw of a card. For example, a Club might indicate a visibility range of 6 inches, a Diamond 9 inches, a Heart 12 inches and a Spade 18 inches.

However, anyone who has tried walking around in fog, heavy rain, or the countryside at night knows that the distance one can see changes quite considerable from moment to moment. Fog swirls, rain arrives in waves, and clouds cover the moon one moment and part to reveal all the next. This can be simulated by drawing a card to fix range of vision at the start of every turn. If one really wants to introduce maximum chaos, simulating a tropical storm at night perhaps, one might draw a card at the start of every player phase.

Random Events

My father was eventually invalided out of World War II when he was wounded during the breakout from Anzio. The Shropshires were advancing across open ground when they were suddenly shelled. He never found out whether he had been hit by German, British or American fire – and it hardly mattered to him at the time. He was rescued by an American Army jeep that was inexplicably driving from enemy-held territory into a British zone.

The point is that completely random but potentially lethal events outside of the control of either side occur in the combat zone. In the twentieth century, these events commonly were shelling, aircraft attacks and mines. And a mine or falling lump of high explosive is on no one's side. Examples for pulp fiction games might include meteorites, volcanic rocks, or bits of a spacecraft disintegrating in the upper atmosphere.

The mechanism for incorporating random destructive events involves setting up an imaginary clock face in the exact centre of the playing area, with north being twelve o'clock. At the start of each turn one player draws a card to give direction from this central point: an ace equals one o'clock, a 5 equals five o'clock, a jack equals 11 o'clock and so on. A king (13) means that no random event takes place. The second player draws two cards and adds the result together to get distance in inches. So a 9, followed by a queen and 3, would indicate that a shell (or whatever) has landed fifteen inches from the centre of the playing area in the direction of nine o'clock, i.e. due west.

Mark the exact spot the shell lands and then work out any effects. Exactly what these effects are is up to the scenario designer but should include impact effect in the form of 'number of attack cards', armour piercing value (if any), and range(s) of effects from the impact point. Some examples below.

Large HE shell
Any model within 3 inches of the impact point is attacked with three cards, Armour Piercing 5.

Any model within 3 to 5 inches of the impact point is attacked with one card, Armour Piercing 3.

Gas Shell
Any model within 1 inch of the impact point is attacked with 1 card, Armour Piercing 1.

Any model within 1 to 5 inches of the impact point is attacked with one card, and if knocked down is 'gassed'. Gassed means that the model recovers at the end of the turn without testing to see if it is killed.

Debris from exploding aircraft/space craft/Zeppelin

Any model within 1 inch of the impact point is attacked with three cards, Armour Piercing 5.

Any model within 1 to 3 inches of the impact point is attacked with one card, Armour Piercing 1.

Interdimensional flux gate

Any model within 1" of the impact point is sucked into the time vortex and is killed immediately, not knocked down.

Any model within 1" to 3" of the impact point is attacked with one card, armour piercing 1

Deteriorating Environments

There are essentially two ways that the environment could deteriorate during play. The first is where something that impacts the performance of models increases step-by-step across the whole playing area at the same time.

For example, battle smoke may thicken turn-by-turn, decreasing visibility. Start with a visibility of, say, 30 inches and then each turn after the first, reduce visibility by 5 inches. If you intend play to go beyond five turns then it is a good idea to stop further reductions in turn six and beyond. This reduction can be randomized to add drama to the scenario. Draw a card at the start of the second turn and thereafter. On a Diamond reduce visibility by 3 inches, a Heart by 6 inches, and a Spade by 9 inches. On the draw of a Club, visibility *improves* by 3 inches as a gust of wind whips some of the smoke away.

The deteriorating condition could be an increase in some airborne pathogen or poison that starts to attack unprotected models when it reaches a certain level, possibly with more dangerous attacks at various trigger levels of intensity. Another possibility is something like flooding, which starts to impact models' mobility.

The second type of environmental deterioration is where something unpleasant starts to creep across the table in a line from one of the non-player table edges to the other. This 'something' could be a rising tide, assuming the battle is taking place in the littoral, or a slow lava flow from some geological event, a carpet of small but loathsome vermin, poisonous snakes and the like, or perhaps a heavier-than-air gas creeping along the ground.

Again the scenario designer has two things to consider: how fast does the effect flow across the playing area, and what does it do? The effect can move at a set pace, say 4 inches per turn after turn one, or excitement can be injected into play by the effect varying in speed. Draw a card at the start of each turn after turn one with, say, a Club equalling no movement, a Diamond 2 inches, a Heart 4 inches and a Spade 6 inches.

The impact of the effect on a model caught in the flow will depend on exactly what the effect is supposed to represent. A lava flow will be immediately lethal to any model, leaving aside pulp-sci-fi weirdness, whereas a semi-harmless creeping gas might attack just unprotected models with only one card. It's all up to the scenario designer.

Conclusion

Skirmish games rely heavily on scenario design and narrative to maintain the interest of players. I have tried to present within these rules something resembling a 'tool kit' with examples, so that players can see how to create and spice up their own stories, based around their own interests and model collections.

But please remember that the game is intended to be played fast – at Hollywood movie speed. All the mechanics are geared to that end and special scenario rules should be no exception.

Notes